abc's
of
Capacitors

by
William F. Mullin

Howard W. Sams & Co., Inc.
4300 WEST 62ND ST. INDIANAPOLIS, INDIANA 46268 USA

International Standard Book Number: 0-672-21498-9
Library of Congress Catalog Card Number: 77-90500

Printed in the United States of America.

PREFACE

Capacitors . . . They have been around since six years *before* Mrs. Franklin told Ben to go fly a kite. Capacitors were discovered quite by accident and regarded as a scientific curiosity for many years. Even today they are still surrounded by much misinformation. For example, how often have you replaced the points and *condenser* in your automobile? The fact is that these devices were originally called condensers. The early scientists thought that capacitors actually *condensed* electricity.

It took quite a few years to dispel this notion. But, it finally became apparent that these devices actually *stored* electrons and that they had a finite *capacity* to do so. In other words, one could only stuff so many electrons into this gadget. . . . That's all. So, the proper name, *capacitor*, came into being. The measure of how much energy one could store came to be called capacitance.

This book is a practical treatise and not a lengthy scientific dissertation. It attempts to answer practical, everyday questions about capacitors and how to use them properly. Yes, there is a certain amount of theory but the reader would not be able to grasp the importance of resonant frequency without some theory. The book will describe in as much detail as necessary how capacitors are made, how they should be used, and, most importantly, why they fail. Practical information on testing and replacement is also included.

After you have digested the contents, you will have a better understanding of the working principles of capacitors. Should you need more detailed information, you will find a capacitor manufacturer ready, willing, and able to provide the information you desire.

This book is actually a compendium of all sorts of information gathered over the past twenty years from a host of sources. Many people provided a great deal of help over the years and the author is sincerely grateful to each and every one of them.

WILLIAM F. MULLIN

CONTENTS

CHAPTER 1

WHAT IS A CAPACITOR?

To the person with even a scant knowledge of electronics, the question, "What is a capacitor?" may seem rather pointless. He probably knows quite well what capacitors look like, having used them in radio and television repair, in experimenting, or in actually constructing various kinds of equipment. But he may never have taken the time to consider seriously the basic principles of capacitors. Does he really know very much about the characteristics, performance, and construction of all of the many types used in present-day electronics?

There are literally dozens of different *kinds* of capacitors, varying in size, shape, and value. They all have one thing in common, yet each has been designed for a specific application where its own particular characteristics are superior to the others. A glance at Fig. 1-1 illustrates the need to explain what a capacitor is and why there is such a large assortment of shapes and sizes.

DEFINITIONS

The simplest definition of a capacitor is: *A capacitor is an electrical device capable of storing electrical energy.* How much energy can be stored and for how long a time is a function of capacitor design.

Fig. 1-1. Common types of capacitors.

Let's look at that definition again. It would be quite possible to confuse this with the definition of a battery. There is, however, an essential difference between the two. A battery is a chemical *generator*. It *produces* electrical energy as a by-product of *chemical* activity. A capacitor is a *storage device*, not a generator; it stores *electrons*. The electrons in a circuit flow into the capacitor and are "trapped" there until they are either removed deliberately, or until they leak out.

Although a capacitor is not a battery, it often behaves as one. In addition, capacitors are capable of acting as resistors, rectifiers, or inductors. Furthermore, they can display all of these characteristics while simultaneously performing the functions of a capacitor.

HISTORY OF THE CAPACITOR

Let's not confuse the issue by jumping into a discussion of the versatility of the capacitor. Let us, rather, go back to the definition of a capacitor—*a device for storing electrical energy.* And let's start at a very logical place—the beginning.

The beginning was in the year 1746 at a place called Leyden, Holland, where a physicist named Pieter van Musschenbrock was performing some experiments in an attempt to "electrify" water. Bear in mind that the year 1746 was in the early dawn of the electrical age. Ben Franklin hadn't even gotten around to flying his famous kite yet. (That happened in 1752.)

At any rate, Pieter van Musschenbrock had a number of crude batteries, a variety of glass jars, and some thin copper foil. He proceeded to line a very thin glass jar inside and out with the copper foil. He then filled the jar with water and placed it inside a larger jar which was also filled with water. When he carefully attached the leads of the batteries to the two separate pieces of foil, nothing happened. Naturally, Pieter assumed that his attempt to "electrify" water was unsuccessful, so he emptied the jars and disconnected the batteries. In so doing, he accidentally touched the leads from the two copper foils and got the shock of his life—literally. This must have been quite a surprise to him, since the batteries were no longer in the circuit. Where, then, did all of this electrical energy come from?

Well, you and I know where the energy came from; van Musschenbrock had, quite by accident, discovered the capacitor. It is interesting to note that the present-day laboratory standard of the capacitor is almost identical to the original device. In fact, it is even called the *Leyden jar,* in honor of its place of discovery. (It is unfortunate that Herr van Musschenbrock's name wasn't shorter and easier to spell, because he gets very little credit for his discovery.)

BASIC CONSTRUCTION OF THE CAPACITOR

The basic construction of the Leyden jar is shown in Fig. 1-2. One copper foil is wrapped around the outside of the glass jar and is connected to the negative terminal of the battery. The second foil is wrapped around the inside of the jar and is connected to the positive terminal of the battery. The two foils are unable to touch each other because they

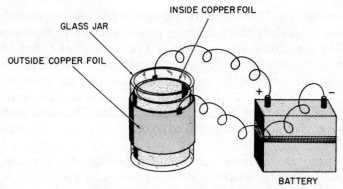

Fig. 1-2. Basic Leyden-jar capacitor.

are separated by the glass jar. Here we have the three essential parts of any capacitor: two conductors (the copper foils) and an insulator (the glass jar). Each of these three parts has a specific name. The conductor (foil) connected to the positive battery terminal is called the *anode*, the conductor (foil) connected to the negative battery terminal is called the *cathode*, and the insulating glass jar is called the *dielectric*. It makes no difference whether the parts are cylindrical (as in the Leyden jar) or flat; the names remain the same—anode, cathode, and dielectric.

Suppose we take the cylindrical Leyden jar and flatten it. The essential ingredients of the capacitor are still the same—two conductors separated by an insulator (dielectric), as shown in Fig. 1-3. If we remove the glass from between the two conductors, we would still have a capacitor. The only difference is that air would become the dielectric. The dielectric can thus be *any* insulating material: glass, air, vacuum, paper, mica, plastic, ceramic, oil, etc. The essential

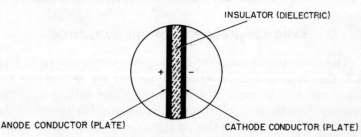

Fig. 1-3. Basic parts of a capacitor.

thing is to be certain that the conductors do not touch, because if they do, the ability of the device to store electrons will be destroyed.

BASIC CAPACITOR THEORY

We have now established that there is no electrical connection between the two conductors (or *plates*, as they are commonly called). Each of the plates is connected to a power source, but the plates are separated. Thus, there should be no current. However, there is a *tendency* for the electrons to flow. The electrons on the negative plate (cathode) *want* to pass through the dielectric to the positive plate (anode). Thus, we can say that a *potential* exists between the two plates and that the electrons tend to accumulate on the surface of the negative plate. This is shown in Fig. 1-4A.

If we disconnect the battery, the electrons will be trapped on the surface of the negative plate (cathode). This is shown in Fig. 1-4B. Now, if we touch the two leads connected to the plates, the electrons will have a way to reach the positive plate. Since electron movement is virtually instantaneous, a spark will be produced (Fig. 1-4C).

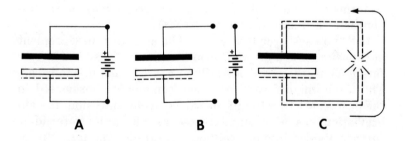

Fig. 1-4. Electron storage on the surface of the negative plate.

The number of electrons stored on the surface of the cathode plate depends on the *area* of the plate and how closely the two plates can be held together without actually touching or providing a conducting path from plate to plate. The amount of voltage being applied has nothing to do with the entire matter so long as the voltage is sufficient to overcome the inherent resistance in the plates and leads. Obviously, the more voltage, the faster the entire affair occurs.

We can now see that our electrical device has a definite *capacity* for storing electrons and that it can be precisely defined. The correct term for this is *capacitance*. Our device, therefore, is called a *capacitor*.

UNITS OF MEASUREMENT

Although Pieter van Musschenbrock discovered the capacitor, another noted scientist's name is more commonly used in connection with the device—Michael Faraday. Faraday developed the method used to measure capacitance. In fact, the basic unit of capacitance—the *farad*—is named after him. A capacitor which, when charged to a potential of one *volt*, carries a charge of one *coulomb*, is said to have a capacitance of one *farad*.

Now, one coulomb is a rather large quantity of electricity, being the amount that passes any point in a circuit when one ampere of current flows for one second. As one might suspect, a one-farad capacitor would be enormous indeed. In addition, potentials on the order of one volt are rather small and impractical. As a result, most capacitors are measured in terms of millionths of a farad (*microfarads*) or in millionths of a microfarad (*picofarads*).

Let's go back over that again. The standard measurement of capacitance is a *farad*. One millionth of a farad is a *microfarad*. Thus, there are one million microfarads in one farad. The term "microfarad" has been commonly abbreviated in the past as MFD, mfd, mf, or μf. To avoid confusion, the abbreviation has been standardized as μF. A microfarad is further divided into one million *picofarads*. The term "picofarad" is rather new and has replaced the older term, "micromicrofarad," which is rather long and unwieldy. Picofarad is shorter and leads to less confusion. Thus the old abbreviations of MMFD, mmfd, mmf, and $\mu\mu$f have been consolidated into a single one—pF. Remember these relationships:

1 farad (F) = 1,000,000 microfarads (μF)
1 microfarad (μF) = 1,000,000 picofarads (pF)

1 picofarad (pF) = 0.000001 microfarads (μF)
1 microfarad (μF) = 0.000001 farad (F)

The common practice is to think in terms of microfarads down to 0.0001 μF (100 pF) and in terms of picofarads up to about 1000 pF (0.001 μF). Note that there is a crossover point. The important thing to remember is that one should use the minimum number of figures. In other words, say it in the quickest and simplest way possible.

VARIATIONS IN CAPACITORS

We have established that capacitance is a function of the size or area of the capacitor plates and the distance between the plates. It follows that we can increase capacitance by either increasing the size of the plates or moving them closer together. Of course if the plates touch the ballgame is over . . . no more capacitor. So, how are we going to keep the plates apart for sure and yet keep them very close together.

In the case of the original Leyden Jar, Herr van Musschenbrock used thin glass to keep the plates separated. As you are well aware, glass is a pretty good insulator. And, as you also are well aware there are insulators, and, there are insulators.

For example, you would be out of your mind to pick up two wires hooked to a 220 volt power line with your bare hands. And a pair of thin cotton gloves wouldn't be much help either. On the other hand you could get the job done with a pair of heavy rubber gloves. The same reasoning applies to the dielectric in a capacitor. Materials with better insulating properties are more capable of preventing electrons from moving from one plate to the next.

The relative quality of the dielectric (insulating) material is called the *dielectric constant*. It's a relative number and has no meaning in and of itself but it does allow convenient comparisons between various dielectric materials. It's all based on the relative insulating qualities of dry air. Dry air has been assigned the dielectric constant of *one* (1). Other materials are either better (or worse) insulators than dry air. A total vacuum also has a dielectric constant of one while

Table 1-1. Approximate Dielectric Constants of Common Capacitor Materials

Material	Typical Dielectric Constants
Air	1.0
Aluminum oxide	10.0
Beeswax	3.0
Cambric (varnished)	4.0
Celluloid	4.0
Glass (Pyrex)	4.2
Glass (window)	7.6
Mica (clear India)	7.5
Mylar	3.0
Paper (Kraft)	4.0
Porcelain	6.2
Quartz	5.0
Tantalum pentoxide	26.0
Vacuum	1.0
Ceramic	12–400,000

tantalum pentoxide is 26. Table 1-1 provides a look at some common capacitor dielectrics.

The quality of the dielectric is crucial to the capacitor because it determines all kinds of things about how the capacitor will operate. We have shown that the plate area and distance between the plates determines the capacitance. The insulating ability of the dielectric, therefore, assumes great importance because the distance between the plates will be determined by the dielectric material.

To illustrate the variations in capacitors, let's compare two of them. One is rated 50 μF at 5 volts while the other is rated 50 μF at 450 volts. What is the difference between them? Why?

In order to understand the difference we must examine the formula used to calculate capacitance:

$$C = 0.2235 \frac{KA}{d} (N - 1)$$

where,

C is the capacitance in picofarads (pF),
K is the dielectric constant of the insulator,
A is the area of one plate in square inches,
d is the distance between plates in inches,
N is the number of plates.

Using the capacitance formula and the values in Table 1-1, we find that two 1-inch square plates, separated by a piece of Kraft paper (K = 4.0) 0.001-inch thick will have a capacitance of 894 pF.

$$C = 0.2235 \frac{KA}{d} (N - 1)$$

$$= 0.2235 \times \frac{4}{0.001} \times 1$$

$$= 0.2235 \times 4000$$

$$= 894 \text{ pF}$$

As you can quickly calculate, using two pieces of one-mil (0.001-inch) paper will cut the capacitance in half. Doubling the area of the plates will double the capacitance. Cutting the thickness of the paper in half will also double the capacitance, but it will also increase the possibility of a voltage flashover.

A voltage of 5 volts is less likely to produce a flashover than is a voltage of 450 volts. Therefore, the principal difference between capacitors of the same capacitance but different voltage ratings lies in the nearness of the plates or the quality (K) of the dielectric.

We can now begin to see the problem facing the capacitor designer. He must constantly compromise size, dielectric quality, and operating voltage requirements. As we progress through the book, you will see that there are also a number of other important factors that must be considered.

CAPACITOR SAFETY

Before we close this chapter, let's take a moment to dwell on the potential danger involved in handling capacitors. One can receive a nasty shock, or even a fatal one, by failing to observe certain elementary safety precautions.

Capacitors can be deadly—don't ever forget that. A current as small as 5 milliamperes can be lethel under certain conditions. Before handling a capacitor, be certain that it has been properly discharged. This does not mean using a screwdriver or touching the leads together. Capacitors should be discharged only through an appropriate resistor. A very rapid discharge through a short conductor can ruin an otherwise valuable component. Also, make certain the discharging resistor is of adequate size, to prevent its being damaged. *Remember:* A capacitor of only 1 μF, charged to 1000 volts, can deliver up to 50 watts of power. So handle capacitors with respect.

CHAPTER 2

CAPACITOR THEORY

In Chapter 1 we established that a capacitor is a device capable of storing electrical energy. Although this is true, a capacitor displays another performance characteristic of great importance in electronic circuits: It blocks direct current but *appears* to pass alternating current. This characteristic is explored fully in a subsequent chapter. We point the fact out here in order to acquaint you with it for future reference.

A theoretically perfect capacitor is described by a symbol (Fig. 2-1A). The equivalent circuit of an actual capacitor (Fig. 2-1B) indicates that there are other characteristics besides capacitance to be considered. There is the inductance (L) of the leads and the plates, the resistance (R1) of the dielectric (leakage resistance), the resistance (R2) of the leads and plates (effective series resistance), and the measured capacitance (C). Each of these factors must be carefully considered when one is using a capacitor in a circuit. For a given application, one type of capacitor having a certain electrical rating may work perfectly, while another of

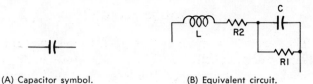

(A) Capacitor symbol. (B) Equivalent circuit.

Fig. 2-1. Symbol and equivalent circuit for a typical capacitor.

exactly the same electrical rating but of a different type and construction may not work at all.

ENERGY ABSORPTION

Whenever a capacitor is charged or discharged, work is performed; in other words, energy must be expended. A certain amount of voltage is required in order to produce enough current to charge a capacitor. The product of voltage in volts and current in amperes is power in watts. The power that we put into a capacitor must eventually be expended in some manner. Most of it will be held in reserve until the capacitor is discharged. The remainder of the power, however, will be dissipated in the form of heat.

A perfect capacitor would store indefinitely all of the power put into it. But there is no such thing as a perfect capacitor. Some of the power is lost in overcoming the resistance (R2 in Fig. 2-1B) of the leads and plates of the unit. More is lost simply because there is no such thing as a perfect insulator (dielectric). Therefore, some of the power is lost in the form of electrons leaking from one plate to the other; symbolized by resistance R1 (Fig. 2-1B). These power losses occur during both the charge and discharge cycles.

The power losses encountered appear as heat in the capacitor. The more rapidly we charge and discharge the capacitor, the more heat we generate. Obviously, these heat losses must be held to some reasonable level. So we have another complication to be considered by the capacitor designer. He must take steps to ensure that the heat will be dissipated in order to avoid damage to the capacitor. The same thing applies to the user. He must be sure he doesn't exceed the critical point where the capacitor can be damaged.

POWER FACTOR

The term "power factor" is used to describe the overall efficiency of a capacitor. It compares the total power required to charge a capacitor with the power lost in doing so. It is a rather complex matter involving all the losses encountered, and it depends on temperature, frequency, dielectric material, voltage stress, and the resistances of the leads and plate material. Although complex, we can describe these factors in relatively simple terms to help you understand their importance.

First, we will consider *frequency*. Everything in nature has a tendency to vibrate, and the vibrations occur more readily at one frequency than another. Witness a window glass rattling as an airplane passes overhead, or a crystal goblet shattering when a certain high note is played on a violin. These are examples of mechanical resonant frequencies. Capacitors are subject to the same sort of resonant frequencies, because even though they are called electrical devices, they are still constructed of mechanical parts which can vibrate physically. Such vibrations can cause variations in capacitance as the distance between the plates varies.

The greater effect in capacitors, however, is electrical in nature. When an inductor and capacitor are placed in series, they form a resonant circuit which has the property of offering a very low opposition to changing (alternating) current at some particular frequency, but greater opposition at all others. The frequency at the point where this opposition is lowest is known as the *resonant frequency* of the circuit.

So how does all this affect capacitors? As the resonant frequency is approached, the apparent capacitance drops off very sharply and the power factor increases rapidly. Here's an example of what we're talking about. Suppose we are replacing a capacitor in a circuit which handles a frequency of 2000 kHz. The original unit was rated at 0.1 μF, 600 Vdc. Can we replace it with one, which just happens to be there on the bench, rated at 0.1 μF, 3000 Vdc? Not necessarily, because the resonant frequency of the original unit was selected to be well above 2000 kHz, whereas the resonant frequency of the second unit could very well be at or near 2000 kHz. Use of the second unit would be wasted effort, because it could quickly overheat and fail.

A second important frequency consideration is lead length. The shorter and thinner the leads are, the higher the resonant frequency becomes. Let's consider another example, involving 5500 MHz. The original unit was rated at 0.01 μF and its leads were one-half of an inch long. It is in a particularly awkward spot, so we'd like to install a replacement unit with 2-inch leads for convenience. Can we do it? Again, the answer is not necessarily. The half-inch leads have a resonant frequency well above the signal frequency in our circuit, but the 2-inch leads may resonate near 5500 MHz. This could result in improper circuit operation, meaning we'd just have to do the job over.

The conscientious technician will always keep all of the above facts in mind and will remember to use replacement capacitors that have, as nearly as possible, the same characteristics as the original units. He will also install the replacements in as nearly the same position as the original units as possible. This is especially true in highly critical circuits and/or circuits where frequencies are relatively high.

TEMPERATURE

Capacitors are affected more by temperature than by any other environmental condition except humidity. This is because capacitors are electrical devices and they store electrons. As temperature increases, electron activity increases, and vice versa. Thus, it can be seen that some design temperature must be chosen to establish a particular capacitor rating. The temperature of 25°C (77°F) is the one most generally used. Any deviation from this temperature will change the capacitance and allowable operating voltage.

As a general rule, capacitance increases as temperature increases, and vice versa. The exact amount of change depends on the type of capacitor in question. There are some exceptions to this rule. Many capacitors, particularly ceramic and polystyrene types, are specifically designed to decrease in value with an increase in temperature. A more detailed account of this effect will be found in Chapter 3.

Temperature has its principal effect on the power factor or overall efficiency of the capacitor. As temperature decreases, internal losses increase, and therefore more potentially destructive heating can occur. Conversely, a noticeable temperature increase breaks down the insulating ability of many dielectrics and thus lowers the rated working voltage of the capacitor.

The effect of temperature on a capacitor is known as its *temperature-coefficient of capacitance*. The actual expression is in terms of percent of capacitance change per degree deviation from a given temperature (usually 25°C). It is further expressed over a given range of temperatures. For example, a capacitor may have a change of +0.01%/°C from +25°C up to +85°C. This capacitor is said to have a positive temperature-coefficient. This is because the capacitance increases as temperature increases. However, many types of capacitors have negative coefficients. Still others exhibit an almost zero change. Fig. 2-2 gives you a general idea of the

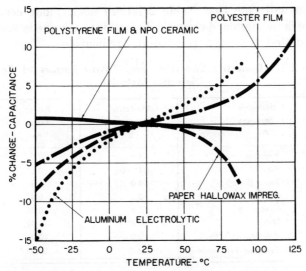

Fig. 2-2. Effect of temperature on different types of capacitors.

effect of temperature on several different types of capacitors. These curves are intended to serve only as an approximation of actual values in order to illustrate the principles involved.

DIELECTRIC MATERIAL

The dielectric material is also affected by a change in temperature. The relative merits of various materials were discussed in Chapter 1, but only from a dielectric or insulating standpoint. Now, let's examine these materials from a power-factor standpoint. It must be emphasized that the values in Table 2-1 are only approximate; they may vary in specific cases.

It is interesting to note that although *Mylar* has only a slightly lower dielectric constant than paper, its power factor is substantially lower. As a result, a *Mylar* capacitor should have lower internal losses than its equivalent paper unit, even though its dielectric might have to be slightly thicker.

The more common dielectrics tend to become less effective as their temperature increases. The lower efficiency adds to the already increased electron activity and in this way it hastens voltage breakdown.

You will note that Table 2-1 does not include ceramics. There is no standard ceramic dielectric; each one is specially formulated to suit a particular application.

Table 2-1. Dielectric Constants and Power Factors of Common Dielectric Materials

Dielectric Material	Dielectric Constant K	Power Factor %
Vacuum	1.0	0
Air	1.0001	0+
Paper (Kraft)	4.0	3.0
Mylar	3.0	0.5
Glass	7.6	0.06
Mica	6.85	0.02
Aluminum Oxide	10.0	10.0
Tantalum Pentoxide	26.0	10.0

DIELECTRIC STRESS

Too high a voltage will weaken the dielectric strength of an insulator. Thus, a dielectric rated at 600 Vdc will probably break down at 1600 Vdc. Furthermore, even though a breakdown does not occur at the higher potential, it will leave the dielectric susceptible to future failure even at its rated voltage.

As would be expected, an important factor involved in dielectric stress is the quality of the dielectric itself. Impurities, especially if they are metallic, can produce areas of poor resistance and can result in breakdown. During the manufacture of paper capacitors, minute amounts of impurities will creep into the fibers of the paper pulp, even with the most careful quality control. To minimize the alignment of these conductive particles, paper capacitors are commonly constructed of many layers of paper.

Dielectric stress is expressed in terms of volts per mil (0.001 inch) thickness—the better the dielectric, the higher the volts per mil. However, this figure will decrease as the temperature, frequency, and/or thickness of the material increase.

PLATE LOSSES

The overall performance of a capacitor depends on a wide variety of factors. Of these, dielectric efficiency is of primary importance, but the efficiency of the plates (conductors) cannot be overlooked. The choice of plate material must of necessity be a compromise. Copper would be the obvious choice electrically, but it corrodes easily and is relatively less stable than aluminum is. Therefore, aluminum has become the most common capacitor plate material in spite of its slightly lower conducting ability. Silver and tantalum are used in applications where their advantages outweigh their additional expense.

The efficiency of plate materials decreases as temperature increases because of added conductive resistance. In addition, the efficiency of the dielectric decreases. Many other factors are involved, but generally, overall losses increase as temperature increases, and vice versa. In spite of this, many capacitors exhibit increased capacitance as temperature increases. But, the overall losses are higher and overall efficiency is lower.

EQUIVALENT CAPACITOR CIRCUIT

Let's return to the equivalent circuit of a capacitor. While the perfect capacitor would have no losses of any kind, and would merely represent an energy potential inserted in any given circuit, an actual unit has several imperfect characteristics.

Every conductor possesses a certain amount of inductance. A conducting wire or the flat plates of a capacitor will exhibit minimum inductance. If we roll the plates into a cylinder, the inductive value will increase. A further increase will result if we roll the plates into a coil—more turns produce a higher inductance.

A perfect capacitor would have an infinite resistance. In actual practice, however, all capacitors have a certain amount of leakage, which an ohmmeter will show as a finite resistance. The amount of this leakage resistance (R1, Fig. 2-1) will vary from one type and value of capacitor to another. It may range from as high as 1000 megohms in one unit, down to as low as 50 kilohms in another unit. Yet both units may be performing their functions satisfactorily. This matter of capacitor leakage is covered more fully in Chapter 6.

The effective series resistance (R2, Fig. 2-1) of a capacitor is actually a combination of the lead and plate losses and is generally much lower than the leakage resistance. It is useful only in calculating the overall efficiency of a capacitor.

CAPACITANCE MEASUREMENT

By definition, a capacitor is a device capable of temporarily storing electrical energy. That is, the device is said to have capacitance.

How does one measure capacitance? One way is to compare the capacitor against a known standard by means of a bridge circuit. Another is to use a wattmeter, which measures the actual amount of energy. The latter is probably the only true means of fulfilling the requirements of the following formula for capacitance:

$$C = \frac{Q}{E}$$

where,

C is the capacitance in farads,
Q is the quantity of charge in coulombs,
E is the voltage across the capacitor plates.

As an example, assume that we charge a capacitor to 500 volts and then discharge it through a sensitive watt/second meter, getting a reading equivalent to one ampere/second of current. Solving the equation, we find that $C = 1/500$, or 0.002 farad (2000 μF).

The time-constant method is another way of measuring capacitance with substantially the same accuracy. A circuit diagram of this method is shown in Fig. 2-3. This admittedly is a rather crude test, since a stopwatch is used, but it serves to illustrate the principle involved. E is the energy source, S

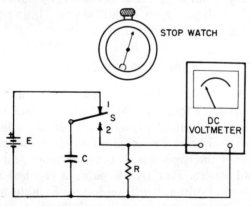

Fig. 2-3. Circuit diagram for the time-constant method of measuring capacitance.

is a single-pole, double-throw switch, R is a 1000-ohm resistor, and C is the capacitor under test. A sensitive electronic (vacuum tube or transistor) voltmeter, with an input resistance of at least 1000 times the value of R, is used for the voltage readings.

While the switch is in position 1, the capacitor remains charged. To start the test, we simultaneously depress the switch to position 2 and start the stopwatch. When the meter reads approximately 37% of the maximum voltage reading, we stop the watch. Capacitance can now be calculated from the following equation:

$$C = \frac{t}{R}$$

where,
C is the capacitance in farads,
t is the time in seconds,
R is the resistance in ohms.

For purposes of illustration, assume that two seconds have elapsed. The equation is solved to give $C = 2/1000$, or 0.002 farad (2000 μF). If the time had been one second, the answer would have been 1000 μF; and so on.

The time-constant test is only as accurate as the timing method used. The one described, using a stopwatch, would be useless for small capacitors. However, by substituting an oscilloscope equipped with a millisecond timing pulse, we could achieve a high degree of accuracy.

CAPACITOR OPERATION

A capacitor does not charge or discharge at a steady rate. At the first instant that voltage is applied, the capacitor will have the least amount of opposition to electron flow. Therefore current will be at a maximum. As electrons accumulate on one plate of the capacitor, it becomes more and more difficult to add others. Finally, the point is reached where the available force (voltage) will no longer be able to increase the charge, and the capacitor will be fully charged. To put it another way, let's go back to the first instant, just as the electrons start to flow. No voltage difference exists across the capacitor (no excess electrons), but as electrons rush to one plate and withdraw from the other, they produce a potential which opposes the applied voltage. There is less voltage difference existing now, so a reduction in current takes place. In other words, the more the capacitor charges, the greater the opposing voltage becomes—until finally it reaches the same value as the applied voltage, and current ceases. The action occurs at an exponential (nonlinear) rate. Fig. 2-4 illustrates the typical response of a capacitor to a charge cycle.

Notice that the current reaches maximum almost instantly, but diminishes gradually to a minimum. It never reaches

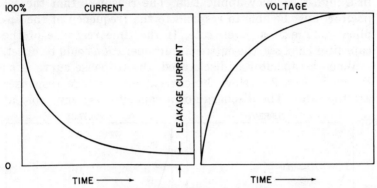

Fig. 2-4. Capacitor charging characteristics.

absolute zero, which is an indication of the leakage current. If the capacitor were perfect, the current *would* drop to zero. The voltage rises in a manner similar to the current decrease, except it finally reaches a maximum value and remains there as long as voltage is applied to the capacitor.

The speed with which a capacitor charges is a function of its size (or capacitance) and the external and internal resistances. If we have two capacitors of equal value, but one has a low internal resistance and the other a high internal resistance, the comparison would be similar to Fig. 2-5.

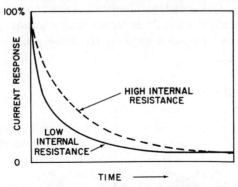

Fig. 2-5. Comparison of charging current responses.

You will note that the current finally decreases to the same level in the higher-resistance capacitor, but takes longer to do so. The actual lag between the two may be only a small fraction of a second, but even this difference can be critical in a high-frequency application. The reason is that the capacitor must be able to respond to the frequency of the applied voltage at a given rate. If the time response of the capacitor changes, the entire circuit operation could be upset.

When a capacitor is discharged, the response curve is as shown in Fig. 2-6. Now the current and voltage responses are the same. The discharge of a capacitor is very rapid at

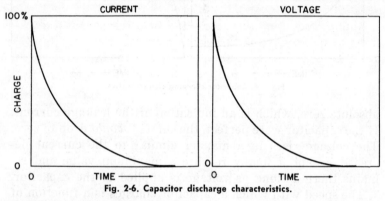

Fig. 2-6. Capacitor discharge characteristics.

first, and then tapers off until the device is apparently completely void of charge.

Perhaps you have experienced this situation and have wondered why: A capacitor, which you are certain has been completely discharged, suddenly shows a charge again. Let's take another look at the discharge characteristic of a capacitor, but this time we'll expand the time scale as shown in Fig. 2-7.

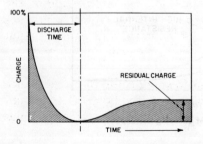

Fig. 2-7. Discharge curve with residual charge shown.

Here we have discharged the capacitor briefly, until it apparently has no charge left. Then we place it to one side. Gradually the charge seems to rebuild until it reaches a definite level. If we now check, we will verify that a small charge has been retained by the capacitor. The reason is that not all of the electrons were evenly redistributed during the discharge operation. This phenomenon is referred to by engineers as *dielectric absorption.* All materials have atomic structures, and all atoms have electrons in orbit around a nucleus. The freedom with which one or more of these electrons enter or leave their orbits determines the relative merit of a material as a conductor.

Application of voltage across the dielectric causes the electrons to be redistributed so that more of them accumulate near the positive plate and fewer near the negative plate. This stressing of the dielectric occurs on initial application of voltage and, in effect, "forms" the dielectric. Never again will the material be the same as it was before the potential was applied. When the capacitor is allowed to discharge, most of the electrons will try to redistribute themselves evenly. But those in the dielectric are "held" fairly tightly, a condition that predominates in insulators. Fig. 2-8 shows the principle involved. A fully charged capacitor is shown as having a dense accumulation (Fig. 2-8A). At the time of discharge, most of them move from the capacitor as shown by the arrow. Some, however, are firmly entrenched and will not leave their orbits (Fig. 2-8B). Thus, it is impossible to ever completely discharge a capacitor once it has been subjected to a dc charge. This condition is depicted in Fig. 2-8C.

TIME CONSTANT

Throughout this chapter, mention has been made of the charge and discharge cycles of a capacitor. In each instance the time has been a factor. This was taken into consideration in the problem of capacitance measurement, but further examination will be helpful.

The formula $C = t/R$, used in solving for the capacitance value, is also the one we will use to explain RC time constants. Expressed in terms of time, the formula becomes

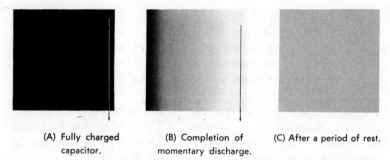

(A) Fully charged capacitor.

(B) Completion of momentary discharge.

(C) After a period of rest.

Fig. 2-8. Illustration of capacitor residual charge.

$t = RC$; t being time in seconds, R resistance in ohms, and C capacitance in farads. The definition of the term "time constant" (1 RC time) is the length of time required to charge a capacitor to 63.2% of the applied voltage. It is further recognized that at the end of 5 RC times, the capacitor will be considered fully charged. Fig. 2-9 shows this graphically.

Since resistance is always present—whether deliberately introduced or contained in the internal structure of the ca-

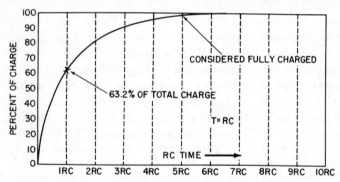

Fig. 2-9. Charge on capacitor at various RC times.

pacitor—the RC time constant must always be considered. The discharge is governed by the same rules as the charging rate. Often the circuit is designed to have a different resistance on the charge than on the discharge cycle. This means the time constants will be different. An example of this is illustrated in Fig. 2-10. The capacitor is charged through a relatively long time constant and discharged through a very short one. When the voltage is first applied to this circuit, the

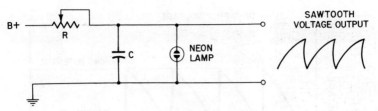

Fig. 2-10. Circuit of neon-lamp sawtooth generator.

potential across the capacitor starts rising from zero, at a rate determined by the values of R and C. If allowed to continue, it would reach the applied voltage in approximately 5 RC times. Before it reaches this point, however, the firing or ionizing potential of the neon lamp is reached and the capacitor now starts discharging through the very low resistance of the ionized neon, meaning the rate will be much more rapid. At some point, determined by the characteristic of the lamp, the voltage will have decreased to the deionizing potential of the neon; the lamp becomes an open circuit, and the capacitor again starts charging through resistance R toward the applied voltage. This action is repeated as long as the voltage is present, producing the series of sawtooth waveforms shown in Fig. 2-11.

The repetition rate, or frequency, of this sawtooth is governed by the values of R and C and by the characteristics of the neon lamp. In the circuit shown in Fig. 2-10 the value of resistor R can be varied to change the frequency.

CAPACITIVE REACTANCE

In the charging of a capacitor, opposition to the flow of electrons is present in the form of resistance and the natural reluctance of the dielectric to accept them. As long as only dc voltages are involved in the charging process, the latter factor is a constant for a given unit. In applying alternating voltages, however, this second factor becomes variable, its value determined by the frequency of the alternations. It is known as *capacitive reactance,* and it opposes the flow of electrons in the charge-discharge process. Like resistance, it is measured in ohms. In applying an ac voltage, a capacitor is first subjected to an attempt at being charged in one direc-

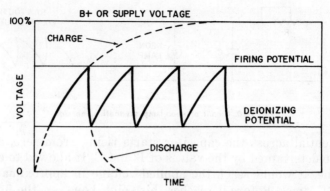

Fig. 2-11. Voltage variation on capacitor in neon sawtooth generator.

tion and then, a short time later, in the opposite direction. This means the capacitor is charged toward a positive potential, discharged back to zero, charged toward an equal negative potential, and again discharged back to zero—this action repeating itself as long as the ac voltage is applied. The length of time the capacitor has to attempt the charge or discharge is determined by the frequency of the applied voltage.

Capacitive reactance is expressed by the following formula:

$$X_C = \frac{1}{2\pi fC}$$

where,

X_C is the capacitive reactance in ohms,
f is the frequency in hertz,
C is the capacitance in farads.

From this formula it can be seen that the higher the frequency and/or capacitance, the lower the capacitive reactance becomes. As the frequency is decreased the reactance increases, until the condition corresponding to zero variation in the applied voltage—in other words, dc voltage—is reached. Here the reactance has reached infinity, as evidenced by the fact that no electrons are flowing because the capacitor has had time to become fully charged.

When frequencies are high with respect to the time constant of a particular RC circuit, a condition will exist where the capacitor will neither charge nor discharge appreciably. In other words, the time for the voltage to change from zero to maximum permits the capacitor to accept only a very few electrons. Therefore, very little of the alternating voltage potential will ever exist across the capacitor. Instead it will appear across the associated resistance. This would be an example of a good coupling circuit (Fig. 4-4) where nearly all the signal voltage will appear across R_g, to be amplified by V2.

IMPEDANCE

In circuits consisting of resistance and capacitance, another term is used to denote the total opposition to electron flow when an ac voltage is present. This is called *impedance*, represented by the symbol Z, and is a combination of resistance R (which remains constant regardless of frequency) and the capacitive reactance (which does not). When the resistance and capacitance are in series, the impedance is computed from the formula $Z = \sqrt{R^2 + X_C^2}$, where Z, R, and X_C are in ohms.

An example utilizing both the impedance and capacitive-reactance formulas will point out the variation existing at two different frequencies. Let us find the impedance of a 100-ohm resistor in series with a 0.1-μF capacitor at 1000 Hz, and then at 10,000 Hz. First we must find the capacitive reactance:

$$X_C = \frac{1}{2\pi fC}$$

At 1000 Hz:

$$X_C = \frac{1}{2 \times 3.14 \times 1000 \times 0.0000001}$$
$$= 1592 \text{ ohms}$$

At 10,000 Hz:

$$X_C = \frac{1}{2 \times 3.14 \times 10,000 \times 0.0000001}$$
$$= 159 \text{ ohms}$$

Notice that X_C is 10 times greater at 1000 Hz than at 10,000 Hz, illustrating that capacitive reactance is inversely proportional to frequency. Now using our impedance formula:

$$Z = \sqrt{100^2 + 1592^2}$$
$$= \sqrt{2,544,464}$$
$$= 1595 \text{ ohms}$$

At 10,000 Hz:

$$Z = \sqrt{100^2 + 159^2}$$

$$= \sqrt{35,281}$$
$$= 188 \text{ ohms}$$

Notice here that the impedance change is not proportional to the change of frequency. This condition can be directly attributed to the addition of the resistive element.

These results show that a different opposition to electron flow exists for each frequency involved.

RESONANCE

Capacitors in combination with inductors form circuits capable of discriminating between different frequencies. These are generally known as *tuned* or *resonant* circuits because they respond only to frequencies in the resonant range. Where a range of selection or rejection is desired, variable capacitors or inductors are employed.

CHAPTER **3**

CAPACITOR CONSTRUCTION

A large variety of capacitor types have evolved from the simple Leyden jar (Fig. 3-1). Some are flat, some are round, some square—in fact, they are to be found in just about every shape imaginable. The reasons for this will become apparent as we investigate capacitor development and construction.

BASIC DEVELOPMENT

Leyden jars were the basic capacitors used in early experiments with electrical energy. Even today they offer several definite advantages. In the first place, they are very dependable. Second, it is relatively easy to change their capacitance by adding or removing metal foil. Third, they introduce very few undesirable properties, such as inductance, into a circuit. However, they are bulky and are limited to comparatively low values of capacitance. This is because a glass jar cannot be made too thin or it becomes too fragile to be handled.

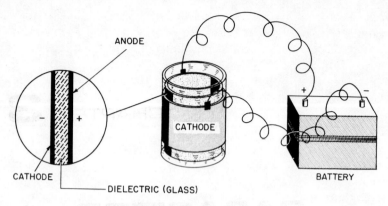

ANODE

CATHODE

CATHODE

DIELECTRIC (GLASS)

BATTERY

Fig. 3-1. Leyden-jar capacitor.

The next logical development in capacitor design was the flat-plate type. It is considerably smaller than the equivalent Leyden jar because there is no wasted space in the center. However, it is still extremely bulky and awkward to handle, especially when larger capacitances are involved.

Of course, flat-plate capacitors can be ganged into parallel circuits, as shown in Fig. 3-2, to provide a very definite saving in space.

Like the original Leyden jar, the first flat-plate capacitors also used a glass dielectric. These units were superior to the

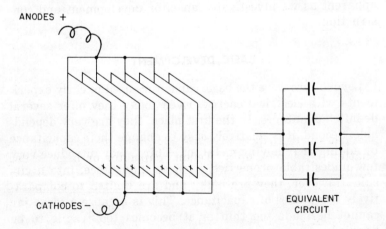

ANODES +

CATHODES −

EQUIVALENT CIRCUIT

Fig. 3-2. Flat paralleled-plate capacitor.

jar, being much more compact; but because of their glass dielectric, they still were quite fragile. It was soon discovered that the glass could be eliminated if the plates were moved farther apart. We now know why—because air can be used as a dielectric; but, since air is less effective than glass, the distance between plates must be made greater in order to provide the same resistance against voltage breakdown.

This is where matters stood until the dawn of radio. Developments then followed in rapid succession—first the mica capacitor, then paper, electrolytics, ceramics, and now plastics.

Until now we've used the terms "glass," "mica," and similar words to describe a particular kind of capacitor. Notice that the term used *generally* describes the dielectric material but not always. As in everything else in electronics, certain slang terms creep into common usage to thoroughly confuse the neophyte. The worst example of this is in the field of integrated circuits but even capacitors have their jargon.

So, let's try to cut through the jargon fog to hang meaningful name tags on various kinds of capacitors. Let's start with the fact that there are only two basic kinds of capacitors: electrostatic or electrolytic.

1. Electrostatic capacitors always have two conducting plates separated by one or more layers of a dielectric material (or combination of dielectric materials as in paper-oil, etc.).

2. Electrolytic capacitors always have but one metallic plate (anode). The second plate (cathode) is always a liquid conductor or a semiconductor solid material.

That's it regardless of what you might read or hear. No matter if you pick up a "Teflon" capacitor or have to deal with the vagaries of a "Computer Grade." The fact is, however, there are a great number of electrostatic types in a wild profusion of sizes and shapes and almost as wide a variety of electrolytics. So let's take a look at the various types of capacitors.

AIR CAPACITORS

Although air is a poor insulator, it is still highly useful as the dielectric in a capacitor because its power factor is almost nil and its stability is excellent. Furthermore, it doesn't cost anything.

The capacitance ratings of air-dielectric capacitors range from about 3 pF, to above 330 pF. Voltage ratings reach a practical limit at about 30,000 Vdc. Air capacitors may be either fixed or variable, but their primary advantage is their relative simplicity.

Fig. 3-3 shows the basic arrangement of a simple air capacitor. The plates of the fixed unit (Fig. 3-3A) are insulated from the supporting frame by an appropriate material. Fig. 3-3B shows a simple, variable air capacitor. A screw thread varies the distance between the two plates and thus

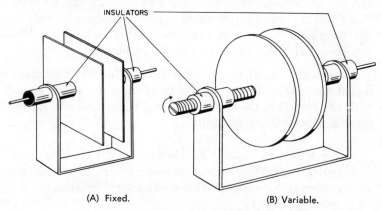

INSULATORS

(A) Fixed.   (B) Variable.

Fig. 3-3. Fixed and variable air-dielectric capacitors.

increases or decreases the capacitance. This arrangement has a disadvantage, however. The resistance to voltage breakdown increases or decreases inversely with capacitance. Therefore, its application is limited.

The more common variable air capacitor is shown in Fig. 3-4. This is the constant-gap type, in which the capacitance is varied by exposing more or less of the plate surfaces to change the area ratio between plates. Since the plates remain the same distance apart at all times, the resistance to voltage breakdown never changes. From this very simple type, the design can be varied in ways limited only by the imagination. Multiple plates may be added to increase the capacitance, and their shape designed to provide the precise variation required. Semicircular plates will provide linear capacitance variation proportional to the amount of rotation. Other shapes can provide a linear frequency variation proportional

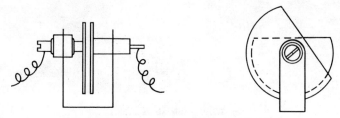

Fig. 3-4. Basic constant-gap variable air capacitor.

to the amount of rotation. Representative types, used principally in tuning and trimming applications, are shown in Fig. 3-5.

The rotor plates can be made of virtually any conducting material, but aluminum is the most common. Other metals such as brass or copper are sometimes used, but these are subject to corrosion unless properly protected by plating. Silver plating is often used to provide a lower surface resistance, and nickel plating is used if extreme corrosion is likely to be encountered.

Although air-dielectric capacitors are among the poorest when judged on the basis of size versus capacitance, they still offer definite advantages over other types. They are extremely stable, are only slightly affected by temperature changes, have a very low power factor, and their life is limited only by physical damage or by failure of the supporting insulators. Their principal disadvantage, other than size, is that they are very susceptible to changes in humidity. This can cause arcing between the plates. Also, the plates may vibrate at high frequencies and cause the capacitance to fluctuate.

In an effort to ward off atmospheric effects, some capacitors are placed in sealed cases which have been evacuated or filled with an inert gas. In some transistor radios they are

Fig. 3-5. Variable air capacitors.

encased in plastic, primarliy to prevent dust particles from forming stray conductive paths and thus affecting circuit performance.

MICA CAPACITORS

Mica was used commercially long before the advent of capacitors. More popularly known as *isinglass,* its most common application was for lanterns, the doors of stoves, etc. This indicates one of the important characteristics of mica: its ability to operate at very high temperatures (up to 500°C). An additional characteristic is that the material is almost totally inert and will not change with age, either chemically or physically.

The development of mica capacitors can be attributed to the fact that mica is usable as a dielectric in its natural state. Mica blocks are mined in conjunction with granite and similar igneous rock formations. The blocks exhibit almost perfect cleavage. That is, they can readily be split into very thin leaves or sheets, often as thin as 0.0001 inch. The sheets are quite uniform in thickness, which is important in capacitor construction. Furthermore the dielectric constant (K) averages about 6.85.

Mica also has some disadvantages. Since it is a natural material, mica is subject to all of the variations of nature.

Great care must be exercised in the selection of mica sheets to ensure that they are as nearly uniform as possible. And, as might be expected, higher-quality mica is also more expensive. As a result, man-made materials often have both a design and an economic advantage.

From a construction standpoint, the mica capacitor is a perfect model of the classic flat-plate type. A thin sheet of mica is sandwiched between two layers of foil (Fig. 3-6).

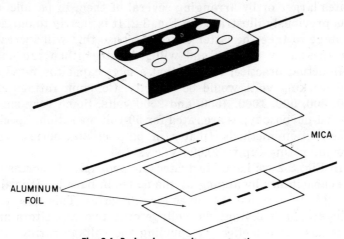

MICA

ALUMINUM
FOIL

Fig. 3-6. Basic mica capacitor construction.

The most common foil material is a tin-lead compound. The tin-lead foil readily conforms to microscopic variations in the surface of the mica. Capacitance is a function of the area of the conductors and the thickness of the mica. It should be noted here that the mica is slightly larger than the foil sheets to prevent contact between the sheets. To show the size/thickness ratio, let us calculate the value of a mica capacitor 1-inch square made from 0.0001-inch mica having a dielectric constant (K) of 7. The formula is:

$$C \ (pF) = 0.2235 \frac{KA}{d} \ (N - 1)$$

$$= 0.2235 \times \frac{7 \times 1}{0.0001} \times (1)$$

$$= 15,645 \ pF \ or \ 0.015 \ (+) \ \mu F$$

Assuming the dielectric strength of mica to be 3000 volts per one-thousandth of an inch (0.001 inch) thickness, our theoretical capacitor has an absolute voltage rating of about 300 volts. In other words, we must never exceed that value. We could therefore safely say that the value of the capacitor would be 0.015 μF @ 150 Vdcw (direct current working volts).

We can increase the value of the capacitor by making the plates larger or by arranging several of them in parallel as was previously illustrated in Fig. 3-2. It is clearly to our advantage to take the latter course because this will increase thickness only slightly and will help strengthen the unit.

In actual practice, we find that a mica capacitor rated at 150 working volts would be a rarity. Common ratings are 100, 300, 500, 1000, 1500, and 2500 volts. Some very small special-purpose types are rated for 50-volt operation. Special construction methods (plates in parallel and series) can provide ratings up to 30,000 Vdcw.

In Fig. 3-2 we have illustrated the principle of increasing the capacitance by arranging the plates in parallel. It is also possible to arrange the capacitors in series. This does two things: (1) it divides the voltage over two capacitors and thus has the net effect of doubling the voltage rating, and (2) it reduces the overall capacitance by the following ratio:

$$C_T = \frac{C1 \times C2}{C1 + C2}$$

where,
C_T is the total capacitance of the two capacitors in series, C1 and C2 are the individual values of the two capacitors.

For example, let us suppose each of the two capacitors has a value of 100 pF @ 500 Vdcw. Thus,

$$C_T = \frac{100 \times 100}{100 + 100}$$

$$= \frac{10,000}{200}$$

$$= 50 \text{ pF}$$

However, the voltage is now divided over two capacitors so that the total rating is 50 pF @ 1000 Vdcw.

Several forms of mica capacitors are shown in Fig. 3-7. These are the classic types of flat-plate units. The interleaved layers of mica and foil are stacked, leads are attached, and the entire structure is dipped in wax to obtain a good moisture seal. Then the assembly is molded into a sturdy block of phenolic resin. The capacitor rating is indicated either by ink stamping or by a color-dot system (see Table 3-1).

Fig. 3-7. Mica capacitors.

As the needs of electronics increased, it became quite apparent that the mica-foil structure needed improvement. Consequently, a method for depositing a thin layer of silver on each side of the sheet of mica was developed. This is done by a type of silk-screen process. The silver is then fired in a furnace. The advantages of this type of mica capacitor over the foil type are outstanding: The silver is in more intimate contact with the mica and thus more potential capacitance is possible; the possibility of air or foreign particles being trapped between the mica and foil is eliminated; physical misalignment between foil and mica is no longer a problem; and, finally, the mechanical resonant frequency of these units is increased because the mica dielectric and the silver conductors are essentially one structure (Fig. 3-8).

The silver-mica capacitor has now completely replaced the old discrete-plate style in all but a very few cases. Indeed, the classic flat molded configuration has now been replaced by modern "dipped" types. The internal construction of the

Table 3-1. Current EIA Standard (RS-153-B) and Military Specification (MIL-C-5C) Color Code for Molded Mica Capacitors

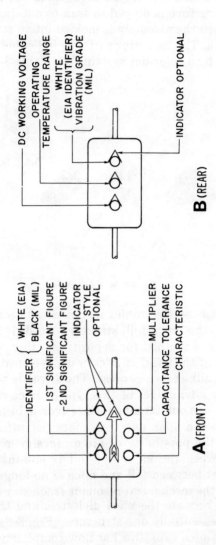

A (FRONT)

IDENTIFIER { WHITE (EIA)
BLACK (MIL)

1ST SIGNIFICANT FIGURE

2ND SIGNIFICANT FIGURE

INDICATOR
STYLE
OPTIONAL

MULTIPLIER

CAPACITANCE TOLERANCE

CHARACTERISTIC

B (REAR)

DC WORKING VOLTAGE

OPERATING
TEMPERATURE RANGE

WHITE
(EIA IDENTIFIER)
VIBRATION GRADE
(MIL)

INDICATOR OPTIONAL

NOTES:
1. The multiplier is the factor by which the two significant figures are multiplied to yield the nominal capacitance.
2. Drawing "A" illustrates standard six-dot system used for "N" temperature range capacitors manufactured according to EIA Standard RS-153-B.
3. Drawings "A" and "B" combined illustrate standard nine-dot system used for "O" temperature range capacitors manufactured according to EIA Standard RS-153-D, and for all units manufactured according to Military Specification MIL-C-5D.

Color	*Characteristic	Capacitance		Capacitance Tolerance	DC Working Voltage	Operating Temperature Range	Vibration Grade (MIL)
		1st and 2nd Significant Figures	Multiplier				
Black		0	1	±20% (M) (EIA)		(M) −55° to +70°C (MIL)	(1) 10-55 Hz
Brown	B	1	10	±1% (F)	100 (EIA)		
Red	C	2	100	±2% (G)		(N) −55° to +85°C	
Orange	D	3	1000		300		
Yellow	E	4	10,000 (EIA)			(O) −55° to +125°C	(3) 10-2000 Hz
Green	F	5		±5% (J)	500		
Blue		6				(P) −55° to +150°C (MIL)	
Purple (violet)		7					
Gray		8					
White		9					
Gold			0.1	±½% (E)† (EIA)	1000 (EIA)		
Silver			0.01 (EIA)	±10% (K)			

* Denotes classification of temperature coefficient of capacitance and capacitance drift requirements.
† Or ± 0.5 pF, whichever is greater. All others are specified tolerance or ± 1.0 pF, whichever is greater.

dipped style is essentially the same as the molded style except that the leads emerge radially rather than axially.

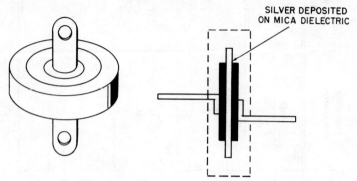

Fig. 3-8. Button style silver-mica capacitor construction.

Basic construction of the dipped-mica capacitor is shown in Fig. 3-9. The silvered-mica sheets are interleaved with connecting lead-tin foils which are folded tightly over the resulting stack. The lead-tin contacts the silvered area on the mica and provides the contact to the clamps which carry the terminal wires. The lead-tin foil is good enough for most applications. However, where performance must be of the very highest order as in military delay lines and critical commercial filters, the lead-tin is replaced by pure silver. Obviously the silver foil increases the price substantially, but it may be the only answer to a critical problem.

After the clamp is applied to the assembly it is immediately given a silicone treatment to eliminate moisture infiltration. It is then dipped in one or more coats of a phenolic material known as *Durez*. The more coats of Durez, the larger the unit becomes. However, the tradeoff is that the more Durez, the more epoxy which can be vacuum impregnated into the final capacitor. It is this epoxy which provides the nearly-hermetic moisture barrier found in the very best dipped mica capacitors.

It is interesting to note, however, that many mica capacitors are sold as "uncased" types. These have only the silicone seal applied and are universally used where subsequent seals will be applied as in special delay lines. These may

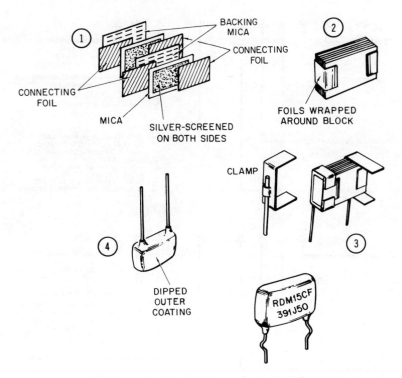

Fig. 3-9. Typical dipped-mica capacitor construction.

contain as many as fifty uncased mica capacitors so it makes sense to avoid the extra size which results from multiple Durez dips.

Standard dipped-mica capacitors are available in capacitance values ranging from 1 pF to 91,000 pF. Style designations are relatively standard in the industry with DM-5 being the smallest and DM-43 being the largest. Many types are covered by both EIA standards and MIL specifications. Type designations are standardized according to the numbering system shown in Fig. 3-10. The style and dimensions of popular silver-mica capacitors are given in Fig. 3-11. MIL specifications now require type-designation marking on both molded- and dipped-mica capacitors. Color coding is now used only for EIA molded-mica types, with type-designation marking optional. All EIA dipped-mica styles are identified by ink stamping; the minimum information consisting of

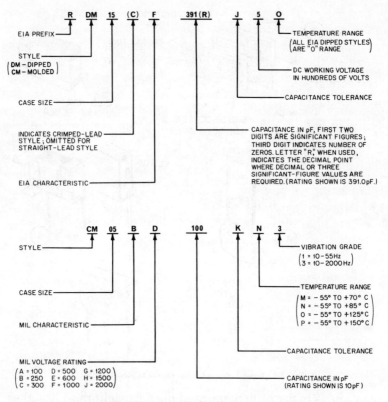

Fig. 3-10. Type-designation numbering system for mica capacitors: EIA at top; MIL at bottom.

manufacturer's name or symbol, nominal capacitance in pF, capacitance tolerance in percent or letter designation, and dc working voltage if other than 500 Vdcw.)

The principal advantage of the mica capacitor is its excellent degree of stability over a wide range of operating temperatures. In addition, mica capacitors are among the best types to use where radio frequencies are involved. However, the performance of the capacitor is directly dependent on the excellence of construction employed. Poorly screened silver, improper clamping and/or dipping will seriously affect performance. Most manufacturers offer "debugged" units which have been operated at full voltage and temperature for 48 hours or more. These units offer superior performance at a modest cost increase.

It is interesting to note that, although mica capacitors are among the oldest of the capacitor types, one seldom sees a microprocessor board without seeing a number of dipped mica capacitors being utilized. The reason is simple. In the ratings used, the mica capacitor offers the best compromise between reliable performance and cost. This is equally true throughout the entire spectrum of capacitors. Because there is no such thing as a "perfect" capacitor, each type and style is the result of a host of performance-availability-cost tradeoffs.

Mica capacitors are relatively bulky when compared to others on a pure capacitance-versus-volume basis. However, from a cost-versus-performance viewpoint, they have few peers.

The principle problem with mica capacitors is that they *are* rather bulky and have specific capacitance limitations. In addition, the flat plates and method of construction can lead to resonant-frequency problems in some circuits. One way to provide the benefits of the mica dielectric is to create a new form of the mica itself. The resulting capacitor is known as a *reconstituted* mica capacitor. Here the mica sheets are ground up and reformed into flat sheets similar to paper with a suitable binder, these flat sheets (or ribbons) retain most of the inherent good dielectric properties and yet are flexible enough to be wound into cylindrical shapes. This accomplishes two things. One, on a rating for rating basis, the reconstituted style is volumetrically more efficient than its flat plate counterpart. Two, it can be more readily encapsulated in standard hermetically sealed metal cases or even molded by the transfer molding process.

But, there's nothing "free" in the world of capacitors. While the reconstituted mica capacitor offers some outstanding advantages, it tends to be substantially more expensive than its dipped competition. This points out once again the compromises that are an inherent part of capacitor selection and use.

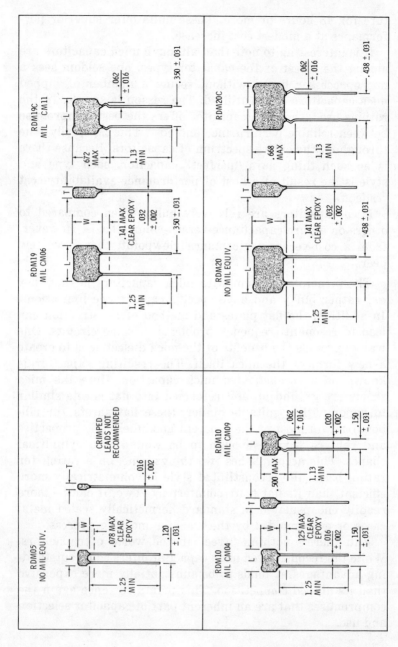

Fig. 3-11. Dipped silver-mica

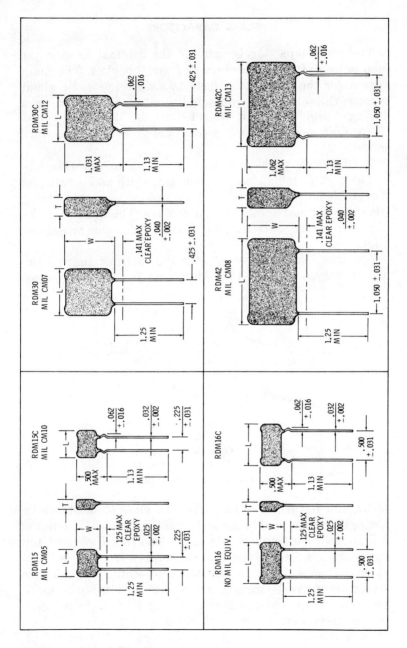

capacitors, styles and dimensions.

GLASS CAPACITORS

The glass capacitor is actually the original Leyden jar modernized after two centuries of much neglect. The main reason for the interest in glass as a dielectric came about because the sources for high-quality mica were endangered during World War II. It wasn't until the 1950s that all of the problems of producing a practical glass dielectric were solved.

The foremost problem was in producing an ultrathin ribbon of highly stable glass of sufficient width and length. As the problems were solved, volume production came about. Glass is superior to mica in many ways. The quality of the dielectric can be controlled, and there are no voids or natural impurities.

Construction is almost exactly the same as that of the mica capacitor (Fig. 3-12). Aluminum foil is used instead

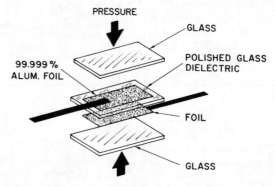

Fig. 3-12. Basic glass capacitor construction.

of tin-lead foil because the surface of glass is substantially flatter than mica, and because the melting point of aluminum foil is much higher than that of tin-lead foil. Layers of foil and glass are interleaved, and the entire structure is fused at high temperature to create an essentially monolithic structure of great strength and superb resistance to moisture. Of the known capacitor types, no other is more resistant to destructive moisture than the glass type.

Capacitance values for glass capacitors range from 0.5 pF to 10,000 pF (0.01 μF), very much the same as mica. Working voltages are comparable to those of mica. By arranging individual capacitors in series and parallel, higher capacitances and very high working voltages are possible. Stability and frequency characteristics are better than those of mica, but cost is substantially higher. As a result, the principal applications for glass capacitors are confined to those areas where improved performance is more important than cost. These applications include air-borne electronic equipment and critical commercial communications gear.

VITREOUS-ENAMEL CAPACITORS

There is a class of capacitors which utilizes vitreous enamel as the dielectric. Their characteristics closely parallel those of mica and glass types. However, the rapid developments in glass have made the vitreous-enamel types virtually obsolete.

PAPER CAPACITORS

While mica and glass have superb qualities as dielectrics, they suffer from a common drawback—they are not sufficiently flexible to be bent or rolled (except for the reconstituted mica described previously). As a result, the use of various types of paper as a dielectric has received a great deal of intensive research and application.

When one thinks of Kraft paper for use as a dielectric, the image of ordinary wrapping paper comes to mind. Nothing could be further from the actual fact. The science of selecting the proper capacitor paper begins in the forest itself. Several species of soft woods are used, and the trees are bred as carefully as modern herds of beef cattle are bred. Each step in processing the wood fiber into eventual paper is subjected to the greatest scrutiny. Many grades of paper are selected depending on the end use. Where the capacitor is to be used primarily with direct current, one grade is used. Where alternating current is involved, other grades are used depending on the amount of ac voltage and current.

The actual construction of paper capacitors varies widely from on class to another, but the basic principle is rather simple. Fig. 3-13 shows that alternate layers of paper and metallic foil are wound into a tight roll. The completed section is vacuum impregnated with a mineral oil (or special synthetic) to improve operating performance. The roll is then suitably encased to prevent moisture infiltration. The selection of case material and the sealing of the case has

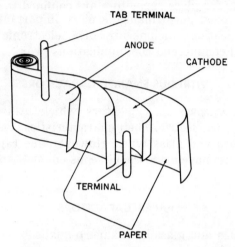

Fig. 3-13. Basic paper capacitor construction.

a tremendous effect on the quality (and life) of a paper capacitor. This is because of the absorption qualities of the paper. Even very minute amounts of moisture gaining access to the paper can have disastrous results. Thus, the highest-quality capacitors use hermetically sealed metal cases, and the lowest quality use a wax-impregnated wrapper of film or paper.

The drawing in Fig. 3-13 shows the standard tab-conductor method of construction. For capacitors carrying large currents there may be several such tabs for each foil. Note that the paper is wider than either of the foils. The disadvantage of this type of construction is that the capacitor will also have inductance; the amount of inductance depends primarily on the number of turns.

It is quite feasible to reduce inductance to near zero. This is accomplished by the extended-foil construction shown in Fig. 3-14. Here, the anode projects beyond the paper on one side, while the cathode projects on the other side. After winding, the respective cathode and anode foils are crushed over the ends of the paper. This effectively shorts the inductance windings without affecting the capacitance. Leads are then either soldered or welded to the crushed ends of the capactor foils. Disadvantages of this system include the fact

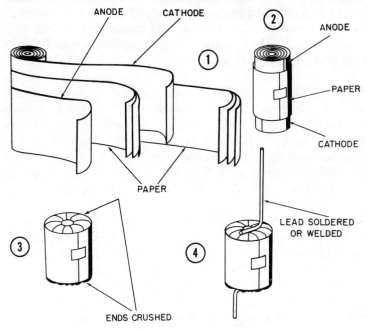

Fig. 3-14. Extended-foil paper capacitor construction.

that the crushing action may warp the roll and degrade the capacitance. Furthermore, great care must be exercised when one is applying the leads, to avoid damage to the conductors or to the paper. A further disadvantage of extended-foil construction is that paper impregnation is more difficult than with the standard tab method. But, as is true in so many other capacitor design problems, one must decide on which factor is more important—lowered inductance or lower cost.

Even though the paper dielectric is very carefully selected in all steps of manufacture, paper is subject to several inherent faults. For example, it is practically impossible to eliminate minute holes in the paper. These holes, of course, degrade the dielectric strength where they occur. In addition, certain impure conducting fibers inevitably appear in the paper. As a result, paper capacitors are manufactured using several layers of paper. This greatly reduces the possibility of holes or conducting impurities lining up. To obtain higher voltage ratings, more layers of paper are generally used. It is also possible to wind the capacitors so that multiple layers of foil and paper are arranged in series for very high voltage applications.

Aluminum foil of very high purity is used for both the anode and the cathode in the vast majority of paper capacitors. Great care is exercised to ensure that the foil is as clean as possible. This is because traces of any contaminants can seriously affect the impregnants used in the final step of manufacture. A wide variety of oil or wax-oil compounds are used as impregnants. The impregnant greatly increases the dielectric strength of the paper and consequently increases the dielectric constant (K).

A further effect of the impregnant is on the temperature characteristics of the resulting capacitor. In the temperature range of from 0° to 85°C, one impregnant provides a positive temperature-coefficient while another provides a negative temperature-coefficient. The actual choice of the impregnant also depends on the ultimate use of the capacitor—one being better for ac while another is better for dc.

Although most paper capacitors use separate sheets of paper and separate layers of aluminum foil (from 0.0015 to 0.001 inch thick) a new type is gaining acceptance. This type is used primarily at low voltages and in solid-state circuits. It is called the *metallized-paper* capacitor. Here, a thin vapor of metal (usually tin) is deposited on the paper and subsequently rolled with other layers of paper. The obvious advantage is that the metal conductor is in much more intimate contact with the paper and thus capacitance is increased. As a general rule, a metallized-paper capacitor carries a higher voltage rating than an equivalent standard unit using the same thickness of paper. This is because any internal arcing caused by voltage stress tends to burn away the very thin layer of metal film. Thus, the capacitor is said to be "self-healing."

The self-healing characteristic can be advantageous in most general-purpose applications (bypass, filter, etc.), but it can be disastrous in a computer RC logic network. Here the signal generated by the internal arcing can completely alter intended circuit signals.

Operating temperatures can radically affect the life of a paper capacitor. For example, a unit designed to operate at 50°C can be expected to have three times as much life if operated at only 40°C. Conversely, operation at 60°C will cut the life in half. This same fact applies to almost all capacitors but it is especially important to the paper type.

The principal advantage of the oil-impregnated paper capacitor lies in its ability to withstand high voltages. Here it is exceeded only by mica, glass, and certain ceramic types. The principal disadvantage is that the upper temperature range is lower than several other types, including film, mica, glass, and ceramic. However, costs are very low for a given amount of capacitance.

PLASTIC-FILM CAPACITORS

Glass capacitors were developed to improve on mica types. The paper capacitor was developed to increase the capacitance available. Now, plastic-film capacitors have appeared on the scene. In fact, the emergence of the various film types is so rapid that any discussion becomes obsolete before it is finished. This portion of the book, therefore, covers only those types in general use at the time of this writing. Some examples of plastic-film capacitors are shown in Fig. 3-15.

Polyester-film capacitors are used in almost every conceivable piece of electronic gear used today. Indeed, the event of the solid-state transistorized circuit has vastly expanded their use. Several manufacturers supply the film in slightly differing compositions: E. I. du Pont de Nemours & Co., Inc., calls theirs *Mylar,* which is a registered trade name; Eastman Kodak Co. calls theirs *Kodar,* which is also a registered trade name. The principal advantages of polyester film are: it is quite dense; it can be manufactured into very thin sheets which are quite stable dimensionally; and it has very high dielectric strength coupled with relatively good insulation resistance.

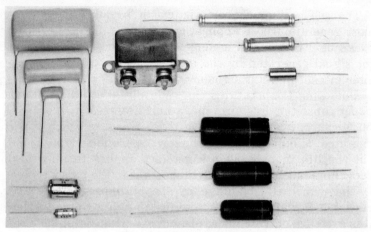

Because of its advantages, polyester film was hailed as the replacement for paper when it first came on the scene. While it is true that for voltages below 600 Vdcw a polyester-film capacitor is substantially smaller than an equivalent paper type, it also exhibits substantial capacitance changes above and below a fairly narrow temperature range. On the other hand, its advantages of size far outweigh its few disadvantages.

Construction of a typical polyester-film capacitor is nearly identical to that of a paper type, the principal difference being that impregnating oils are unnecessary—the dielectric qualities are solely dependent on the film itself. Both tab and extended-foil construction are used, depending on cost and application.

Polyester-film capacitors are supplied in several configurations. The most popular are the dipped and molded styles. Another style which uses an outer film wrap with an epoxy end-fill represents a marked reduction in size. A new style which is gaining in favor makes use of the inherent self-sealing characteristics of polyester film. This new style uses extra wraps of the dielectric itself to form the outer wrap. Wax impregnation and end heat-sealing complete the job. Terminal wires are inserted under heat and pressure to contact the active internal foils. The self-sealed style is the smallest size available. One manufacturer supplies radial-lead types while another uses axial leads. The self-sealed types are of extended-dielectric construction and have some inherent inductance. The dipped, molded, and wrap-and-fill styles are all extended-foil construction with consequent less inductance. For some special military uses metal cases are supplied, but these offer only slightly better long-range performance at vastly increased cost.

Just as there are metallized-paper capacitors, there are also metallized polyester-film capacitors which are commonly called *metallized Mylar,* in conjunction with du Pont's trade name. As in the paper version, metallized Mylar offers tremendous reductions in size because of the extremely intimate contact between the vaporized metal (usually tin) and the film. Indeed, a further development in thin-film polyester and aluminum metallizing has resulted in capacitors of extremely small size and excellent reliability. The principal gain is in volume; however, metallized-film capacitors have the same "self-healing" characteristic as metallized paper. This characteristic is an advantage for many applications, but is a disadvantage for others, such as in computer logic circuits.

Polystyrene-film capacitors offer outstanding stability in a relatively small size. Their insulation resistance is exceptionally high in comparison with all the other commonly used capacitors. The net result of this is that their internal losses are lower than for virtually any other type. Another advantage of polystyrene film is that stability versus time is nearly constant. In other words, age has practically no effect on the capacitance value. A capacitor having a rating of 100 pF will be within 0.1% of that value twenty years later.

A further advantage of polystyrene-film capacitors is that they have a slightly negative temperature-coefficient. Also, polystyrene is relatively simple to work with, and its raw materials are comparatively cheap. Thus, we have available for general use a very high quality capacitor at relatively low cost.

Polystyrene, however, does have some disadvantages. Its upper temperature limit is about +85°C. Above that point the polystyrene starts to deform and the capacitor is destroyed or permanently damaged. The other disadvantage is size. For equivalent capacitance values, these units are larger than polyester types. As a result, typical values for polystyrene types range from 5 pF to 0.01 μF, whereas values for polyester types range from 0.001 μF to 1.0 μF (metallized types extend to 15 μF). A comparison of the two types at 0.01 μF, 600 Vdcw shows the wrapped-polyster type to be about 25% smaller than the equivalent polystyrene type.

Construction of the polystyrene capacitor is unique. (Fig. 3-16). The aluminum foil is narrower than the polystyrene

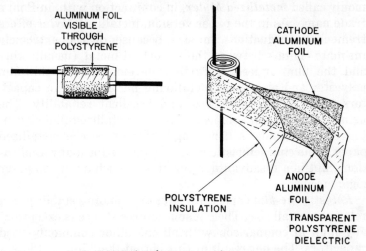

Fig. 3-16. Basic polystyrene capacitor construction.

film. After the capacitor is wound, it is fused so that the extended edges of the polystyrene film provide the moisture barrier. Because of this, extended-foil construction is not commonly employed. It is possible to use extended-foil construction, but then an outer encapsulating case is needed, which adds to the cost and provides only slightly better performance.

Polyester/polystyrene dual-dielectric capacitors have found limited applications. By combining the two films, the best characteristics of both tend to negate their individual disadvantages. These units are smaller than pure polystyrene capacitors, while their stability is much better than pure polyester types. Temperature characteristics provide a nearly flat curve with the negative temperature-coefficient of the polystyrene balancing the positive temperature-coefficient of the polyester.

Polycarbonate capacitors combine nearly all of the good characteristics of both polyester and polystyrene. Temperatures range up to 140°C. Stability is nearly as good as polystyrene. The most unusual characteristic is the temperature-versus-capacitance curve. Polycarbonate has a negative temperature-coefficient above +25°C, and a positive temperature- coefficient below that point. Furthermore, the temperature-coefficient curve is very flat with only a slight deviation. Since polycarbonate film can be made thinner than polyester, it offers marked size advantages. Construction is similar to that of polyester types.

A principal advantage of all plastic film capacitors is that they do not require impregnating oils to provide high voltage operation. For many years the relative costs of plastic film capacitors versus oil-impregnated paper types were in favor of the older oil-paper units. Now, however, environmental considerations have become a critical factor. This is because most common impregnating oils are highly toxic and cause permanent damage to fish and other environmental factors. While it is true that a great deal of work has been done to develop a nonharmful impregnating oil, the relative cost disadvantage of the plastic film capacitor has largely disappeared.

Polycarbonate film capacitors offer some distinct advantages over other film types where high voltage corona is a factor as in SCR (Silicon Controlled Rectifier) circuits. The polycarbonate capacitor offers virtually the same corona resistance as the best paper-oil types in a substantially smaller volume and weight. Indeed, within the past several years metallized polycarbonate has become the standard type specified for SCR commutation and in a wide variety of high voltage ac applications.

Other plastic films include *cellulose triacetate* (CTA), *polypropylene* (PPP), *polyparaxylene* (PPX), *polypyromellitimide* (PPM), and *polytetrafluoroethylene* (PTFE). (It seems that as capacitor technology advances, the names of the materials become more difficult to pronounce). The cellulose (CTA) film is the oldest of the group. However, it offers no important advantage over polyester and will probably never receive wide acceptance.

The other four materials are relatively new. Of these, PTFE (known popularly as *Teflon,* another registered trade name of E. I. du Pont de Nemours & Co., Inc.) offers exciting possibilities for high-temperature capacitors. Its insulation resistance is higher than polystyrene and temperature-coefficient characteristics are very similar. Thus, it would seem to be a better material. Unfortunately, a completed PTFE capacitor costs nearly five times as much as an equivalent polystyrene type.

The constant quest for higher performance capacitors has led to another new dielectric material. This type is known as *polysulfone.* According to manufacturers specifications the polysulfone offers size, stability, and temperature advantages over other film types. Whether or not these claims are of sufficient economic or performance value remains to be seen.

CERAMIC CAPACITORS

No other dielectric is as versatile as ceramic. This manmade material can be mixed in many different ways to produce a wide variety of results. One mixture yields a capacitor with characteristics nearly the same as mica. On the other hand, another mixture produces paperlike characteristics.

In configuration, ceramic capacitors are generally either disc-shaped, tubular, or rectangular. The rectangular configuration is widely used in solid-state electronics, particularly in the newer *monolithic* ceramics which will be discussed in more detail later on in this chapter. Both tubular and disc types are made in adjustable as well as in fixed capacitance values. Fig. 3-17 illustrates several common types.

Capacitance values range from as low as 0.5 pF to as high as 3 μF. Operating voltages vary from a mere 3 Vdcw to over 30,000 Vdcw. Ceramics offer good size-to-capacitance ratios and outstanding temperature and frequency characteristics.

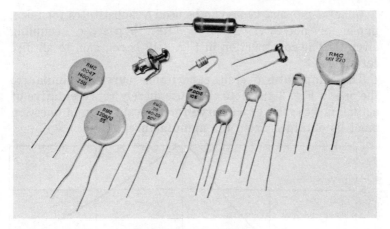

Fig. 3-17. Ceramic capacitors.

Ceramic capacitors have similar capacitance values but lower inductances than mica types of corresponding values. The four types are general-purpose, temperature-compensating, temperature-stable, and frequency-stable. Each of these four broad types has characteristics which suit it to particular applications.

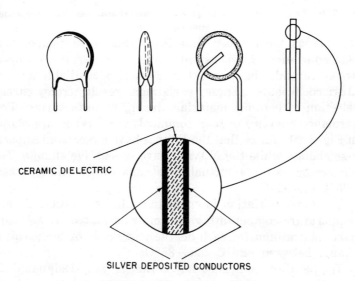

CERAMIC DIELECTRIC

SILVER DEPOSITED CONDUCTORS

Fig. 3-18. Disc ceramic capacitor construction.

General-purpose types can be used as substitutes for mica, paper, or polyester types in filter, bypass, or coupling circuits. The construction in Fig. 3-18 is common to all disc ceramics.

The temperature-versus-capacitance curve is similar to the one in Fig. 3-19. Note that it is merely representative of actual curves; the exact curve for the capacitor in question must be obtained from the manufacturer.

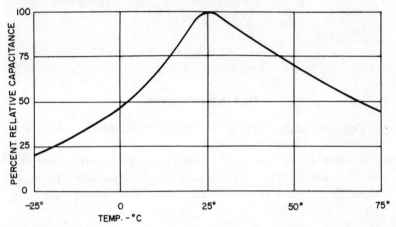

Fig. 3-19. Characteristic curves of general-purpose ceramic capacitors—temperature versus capacitance.

Whereas most other capacitor types have temperature characteristics of a constant variety, *temperature-compensating* ceramics have an extreme range of available temperature-coefficients. These variations result from careful selection of ceramic materials during manufacturing. Performance curves for four commonly used types are shown in Fig. 3-20. Notice that the P types have a positive temperature change while the N types have a negative change. The *NPO* type exhibits virtually no capacitance change from $-25°C$ to $+85°C$.

The exact variation in capacitance to be expected from a temperature-compensating capacitor is stated as so many parts per million for each degree centigrade of temperature change between $+25°C$ and $+85°C$.

Temperature-compensating ceramics are designated by a number with a letter prefix. The system is actually fairly

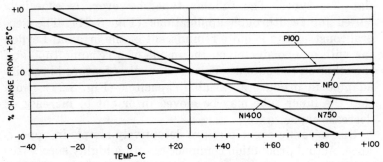

Fig. 3-20. Characteristic curves of temperature-compensating ceramic capacitors.

simple, once you become familiar with it. The prefix *P* means that above +25°C, the capacitance will rise with the temperature (or a positive coefficient); *N* means the capacitance will drop (negative coefficient); and *NPO* signifies no change. The larger the number following the prefix, the greater the change. *NPO* types will, of course, have no number.

Thus, a designation of P100 means that for every degree centigrade that the temperature increases (between +25°C and +85°C), the capacitance will rise 100 parts per million. An N750 rating produces a decrease of 750 parts per million per degree increase, whereas an N1400 produces almost twice that amount.

These capacitors are particularly useful in compensating for value changes which occur in other components when equipment temperature rises.

Temperature-stable ceramic capacitors are essentially refinements of the *NPO* temperature-compensating types. Their temperature range is extended so that from −60°C to +100°C, the capacitance changes only ±7.5% from the stated value at +25°C.

Frequency-stable ceramics are compounded so that they maintain a relatively constant resonant frequency over their operating temperature range. In contrast, temperature change has a very marked effect on the resonant frequency in standard ceramics.

Tubular ceramic types have a poorer size-to-capacitance ratio and slightly more inductance per microfarad than the flat disc types. The tubular type, however, is especially suited for feedthrough applications (Fig. 3-21). The general construction is quite reminiscent of the original Leyden jar with one conductor on the inside of a ceramic tube and the other on the outside. This construction feature is readily adapted to produce a variable capacitor (Fig. 3-22). Note that the inner tube may be moved in or out to increase or decrease capacitance.

Disc ceramic capacitors are deceptively simple in appearance. Actual production commences with a highly specialized

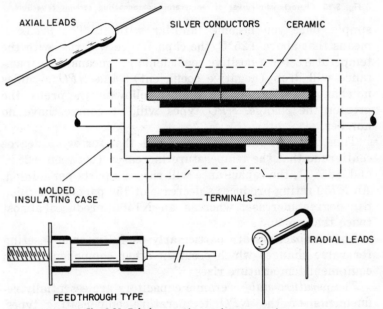

Fig. 3-21. Tubular ceramic capacitor construction.

and formulated ceramic powder. This powder is mixed with a resin and the mixture is pressed into a disc under high pressure. The disc is then fired in a furnace at high temperature to form a rugged base for the conductors. The conductors are actually pure silver paint applied to both sides of the disc by a silk-screen process. After appropriate drying, the assembly is dipped in solder. The solder adheres to the silver but not to the ceramic. After the leads are attached, a tough *Durez* coating is applied for environmental

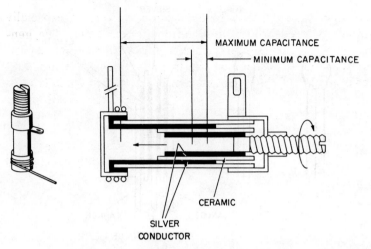

Fig. 3-22. Variable tubular ceramic capacitor construction.

protection. The most common disc ceramic is a single unit. However, two discs may be joined to produce a relatively small unit of up to 0.01 μF rated at over 1000 Vdcw (Fig. 3-23).

Because of their superior dielectric properties, ceramic capacitors are becoming more popular than the older mica types in trimmer applications. There are two common basic configurations. One configuration is essentially a concentric tubular unit, and the other is similar in principle to the variable air type.

The variable tubular type (Fig. 3-22) consists of a silvered supporting tube with one plate on the inside. A movable ceramic tube, fired with silver on its inside, fits precisely within the first tube. As the inner tube is moved, the effective area of the plates changes and varies the capacitance. A typical adjustment range is from 1 to 8 pF.

Another variable ceramic, quite popular as a trimmer, consists of two partially silvered, ground ceramic discs. As the one disc is rotated, it presents more or less effective plate area to the fixed disc. This is very similar to the action of a variable air capacitor, as shown in Fig. 3-24. Its principal advantages are that a 180° rotation produces maximum capacitance change, and that it is available in larger values

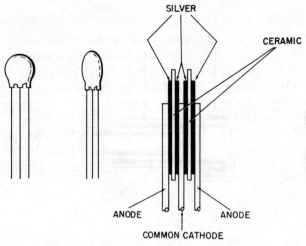

Fig. 3-23. Dual disc ceramic capacitor construction.

than the tubular type. Furthermore, temperature-compensating types ranging from *NPO* through N750 are possible, with ratings of 1.5–7 pF and 8–50 pF. Side-by-side dual-variable types are also available.

Ceramic capacitors of the so-called doorknob type are common in television high-voltage applications. They are supplied in large molded cases with rings arranged to minimize high-voltage flashover.

It is interesting to note that ceramic capacitors compete almost directly with polystyrene-film types in relative size,

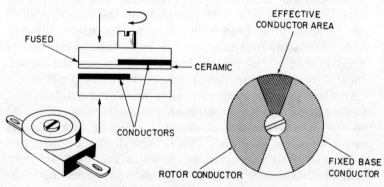

Fig. 3-24. Variable disc ceramic capacitor construction.

cost, and performance. In the United States, the ceramic types are still more popular and are widely used. In Europe, however, nearly the opposite is true, with the polystyrene types being more popular. The Japanese incorporate both types about equally, so that ceramics are used where their performance is best and polystyrene types are used where their performance is best. The service technician should be aware of the interchangeability of the two types.

A slight disadvantage of the ceramic capacitor is its relative fragility. One should avoid exerting more than a minimum amount of pressure on the leads during installation. Excessive pressure can crack the protective coating or permanently damage the capacitor internally.

A recent development in ceramic capacitors has been the emergence of the high-capacitance, low-voltage types. These are marketed under various trade names such as *Hypercon, Magnacap, Transcap, Ultra-Kap,* etc. These types are widely used in solid-state circuits. They compete directly with film and electrolytic types in low-impedance circuits, and offer some size or cost advantages over the other types.

MONOLITHIC CERAMIC CAPACITORS

The advent of solid-state electronics, and particularly integrated circuits, has lead to the development of a new style of ceramic capacitor. This new style combines the basic science of the older disc style with the newer technologies of integrated circuits.

From an electrical standpoint, the monolithic ceramic capacitor is essentially the same as the disc style. Ceramic material is mixed to provide the precise degree of temperature or frequency characteristics needed. The basic difference is in volumetric efficiency. Where a typical disc might offer 10 μF per cubic inch, modern monolithics exceed 1000 μF per cubic inch.

Construction is similar to disc styles in many ways (Fig. 3-25). Silver is applied to thin sheets (ribbons) of ceramic base material; however, the thickness of the sheet is dramatically less in the case of monolithics (.0008 in. versus .032 in.). Multiple layers of silvered ceramic ribbons are

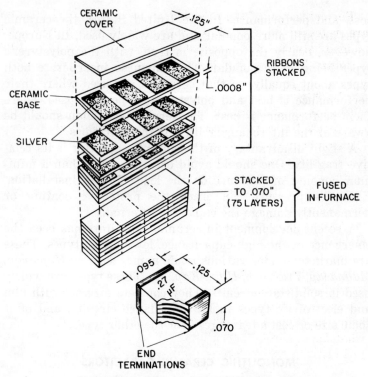

Fig. 3-25. Monolithic ceramic capacitor construction.

stacked and fused in a furnace to form a "block." The block is then cut into individual capacitors. End terminations are applied by means of a metallic ink (silver, palladium, gold, platinum, etc.). Actual terminal wires are then attached to the ink by soldering.

Completed monolithic ceramics are *very* small. A typical .22-μF unit measures a mere .170-in. long by .060-in. wide and .060-in. thick. This extremely small size makes the monolithic ceramic capacitor a natural for inclusion in hybrid solid-state (integrated) circuit packages, and this is the principal use of the product. On the other hand, millions of the devices are being used in dipped or molded form to replace larger disc ceramic, mica, and plastic-film capacitors. As production increases and prices become more competitive, additional millions of units will be sold and used.

For convenience, the monolithics used in hybrid circuits are referred to as "chip" style, while the dipped and molded styles are referred to simply as "monolithic ceramics."

Silver Migration

The extremely small size of monolithic ceramic capacitors has focused attention on a problem common to mica, glass, vitreous enamel, and normal ceramic capacitors. This problem is known as "silver migration." Wherever silver is used in an electronic circuit with a potential applied, it tends to "migrate" to stabilize the electrical imbalance. This migration occurs with startling speed. A thin microscopic tendril of silver will bridge a gap of .150 inch in 103 seconds if moisture is present. This tiny tendril will positively short-circuit a capacitor.

It thus becomes obvious that extreme care must be taken to exclude moisture from capacitors utilizing silver as the conducting plates. An understanding of this phenomenon helps to explain certain sudden, catastrophic failures of what appear to be perfectly good capacitors. It also serves to illustrate the problems facing capacitor designers.

ELECTROLYTIC CAPACITORS

Electrolytic capacitors provide more capacitance for their size than any other type. From the service technician's viewpoint, they are also the most confusing. They seem to defy all the rules applicable to capacitors, yet electrolytics are able to perform jobs which no other type can. Fig. 3-26 shows only a few of the wide range of values and sizes that are available.

Electrolytics all have one thing in common, however—they are made differently from other capacitors (see Fig. 3-27). Instead of the usual plates separated by a dielectric, the electrolytic has a metallic anode coated with an oxide film. This outer covering is the dielectric, and a liquid electrolyte acts as the cathode. A second metallic conductor serves primarily as the connection to the liquid cathode, providing an external termination.

Fig. 3-26. Typical electrolytic capacitors.

In actual practice, porous paper is wrapped around the anode and saturated with the electrolyte to eliminate the spillage problem.

There are two common types of electrolytic capacitors: aluminum and tantalum. Both employ the same basic principles. The aluminum or tantalum anode is covered with an oxide film. A suitable liquid (or solid) electrolyte is the cathode. The aluminum type is by far the most popular because of its lower cost.

The aluminum-oxide film has a very high resistance to current in one direction and a very low resistance in the opposite direction. In other words, the film acts as a dielectric in the first instance and as a plate in the second. Because of this, electrolytics are polarized. If the designated polarity is not observed, the oxide film on the anode will break down and migrate to the cathode connection, resulting in prompt failure of the capacitor.

Electrolytics are described as belonging to one of three basic families (Fig. 3-28). The standard *polarized* type (Fig. 3-28A) has one anode, the liquid cathode, and a cathode

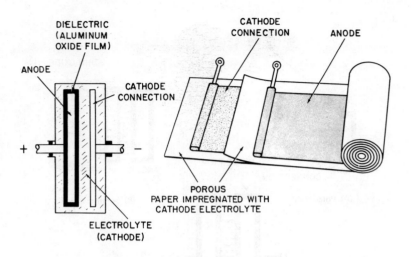

DIELECTRIC (ALUMINUM OXIDE FILM)

ANODE

CATHODE CONNECTION

CATHODE CONNECTION

ANODE

POROUS PAPER IMPREGNATED WITH CATHODE ELECTROLYTE

ELECTROLYTE (CATHODE)

Fig. 3-27. Basic electrolytic capacitor construction.

connection. Polarity must be observed. However, in many energy-storage applications a certain amount of current reversal is often encountered. In this case, the use of a *semipolarized* type (Fig. 3-28B) is recommended. In this type, the primary anode has a relatively thick oxide surface. The cathode connection now becomes the secondary anode with a thin oxide surface. Also, the liquid cathode has a slightly different chemical composition. The third type (Fig. 3-28C) is used in audio crossover networks and ac motor-starting applications. Here, there is a complete reversal of polarity; therefore, two anodes are required. The cathode connection now becomes a second, and equal, anode. Obviously, size will be affected. In fact, this third type (called *nonpolarized*) is just twice as large as a polarized type of equivalent capacitance and voltage rating.

The anodes in an aluminum electrolytic capacitor are radically different from those used in other capacitor types. Rather than the smooth shiny surface one associates with a typical paper or film capacitor, the anode of an aluminum electrolytic appears dull gray in color and has a mysterious brittle feel. The reason is at once simple to express and incredibly difficult to comprehend. In the past it has been commonplace to dismiss these anodes as being somehow

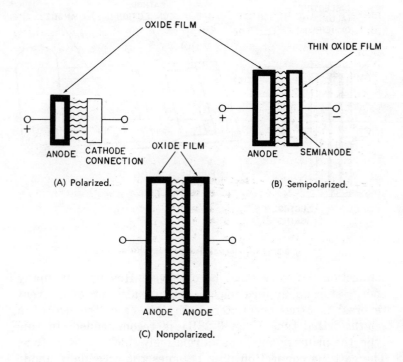

Fig. 3-28. Three types of electrolytic capacitors showing relative size for the same capacitance value.

"roughened" to increase the surface area. Only recently have the true facts become known. It is to the credit of the General Electric Company that time and money was invested to determine exactly what happens during this mysterious "roughening" process.

It was found that during the acid etching process literally billions of tiny tubes were formed in the aluminum foil structure. The actual surface area of the aluminum surface was *not* increased; if anything it was *decreased*. But, what happened was that the surface area available to the subsequent electrolyte was increased by a factor of *fifty* or more.

Consider this analogy. Take a flat plate of aluminum a foot square and roughly $\frac{5}{16}$ inch thick. Now start to drill tiny $\frac{1}{64}$ inch diameter holes about $\frac{1}{64}$ inch apart in the surface of the aluminum. Don't drill deeper than $\frac{5}{32}$ inch and make sure the holes are in a random pattern. Now turn the plate over and repeat the drilling from the other side making certain the holes don't touch the holes on the opposite side. Now shrink the piece of aluminum to one square inch and .002 inch thick.

Does that sound nearly impossible? If not, then this next requirement should boggle your mind. Do the job in *30 to 45 seconds!* That, in essence, is what happens when aluminum foil is etched as the first step in making an aluminum electrolytic capacitor. There is one more interesting fact. Each of these billions of tiny tubes *always* enters the surface at precisely 90° and descends into the aluminum in a series of nonconnected paths that may take as many as a dozen 90° turns. The scanning electron microscope provides fascinating pictures of this scientific fact.

It was found, for example, that it was possible to control not only the size of the holes, but also their distance apart and depths. The practical application of this knowledge came as it was found that low-voltage electrolytics worked better with small closely spaced holes while high-voltage (450 volts) types needed larger more widely spaced holes. The result is that the electronic industry gained new aluminum electrolytics that not only worked *better,* they lasted longer, and cost less.

The aluminum electrolytic capacitor is the same as any other capacitor in that it has an anode, a cathode, and a dielectric. But, just as the anode described previously is radically different than the anode used in other capacitors, so is the cathode. The cathode in a normal electrolytic is a *liquid.* It's a conducting liquid. In this description it is not important to get into the vast varieties of *electrolytes* used to make the cathodes. All that is needed here is the realization that the cathode is indeed a liquid, and that it is basically boric acid and glycol.

Having covered the anode and the cathode let's examine the third unusual part of an aluminum electrolytic capacitor, the dielectric. It's a form of aluminum oxide. This oxide layer is formed on the entire surface of the anode by electrolytic action. The very thin film forms on all parts of the anode. Down into the tiny holes covering everything with one of the best insulators known to science—aluminum oxide. A combination of factors provide the outstanding advantage of high capacitance per unit volume that the aluminum electrolytic capacitor has over other types. That is, the exceptionally vast area of the anode coupled with the highly efficient dielectric plus the intimate contact of the dielectric with the anode and the liquid cathode which permeates every tiny tube.

In actual practice the etched and formed aluminum anode is wound in combination with multilayered paper and an aluminum cathodic *connector* to form a cylinder. This cylinder is then impregnated with the electrolyte which is the actual cathode. Excess electrolyte is drawn off by means of a vacuum or centrifuge to produce a capacitor section which is "dry" in appearance. But, it isn't dry. If it were dry, it wouldn't work. As shown in Fig. 3-27 there are two electrical connections to a wound aluminum electrolytic the tab going to the anode and another going to the cathode connector.

Since it's not practical to use the capacitor in its rough wound form, it is placed in a suitable container of plastic or metal. It's interesting to note here that the old cardboard tube-style container sealed at the ends with wax has virtually disappeared. The reason for the disappearance is simple; the cardboard tubes became more expensive than aluminum cans.

It is often assumed that aluminum electrolytics are "passive" components in contrast to "active" transistors. The true fact is that aluminum electrolytics are far from "passive." All electrostatic capacitors (paper, film, mica, etc.) are essentially "passive" in nature. This is because unless something happens to harm them electrically, they will operate indefinitely. Such is not the case with aluminum electrolytics. The interaction of the voltage applied to the electrolyte and the aluminum oxide causes minute amounts of gas to form. Indeed, as the voltage and current is increased, the amount of gas increases. If the voltage and current increases rapidly and exceeds the rated operating value of the capacitor, the gas increases at such a dramatic rate that the capacitor literally explodes. As a result, most electrolytics incorporate some sort of "vent." While the vent is essentially a safety device, it nonetheless allows a certain amount of gas to escape over the life of the capacitor. This escaping gas represents in a very real sense a loss of electrolyte. Since the electrolyte is actually the cathode itself, a loss of electrolyte means a decrease in the value of the capacitor. In time the capacitor will fail. If operated within its limits this may be ten to twenty years or even more. But, the fact remains that an aluminum electrolytic capacitor will eventually wear out and must be replaced.

It was previously thought that operation of a 450-Vdcw electrolytic at lower operating voltages would harm the unit and result in premature failure. This is not true. While it is true that the 450-volt unit will not perform as efficiently at 150 volts (for example), it will still not harm the unit. However, the capacitor will be larger than necessary and leakage currents will be higher. In case of necessity, it is perfectly permissible to use an electrolytic of higher voltage rating than the original. However, the opposite is *not* true. *Never* make a substitution with a capacitor of *lower* voltage rating.

Electrolytic capacitors have a characteristically higher leakage current than other types. This is due partially to their design and partially to impurities in the foil and electrolyte. Leakage current increases as the temperature rises; simultaneously, the voltage-breakdown resistance decreases. This is a sort of self-feeding situation where the more the leakage, the hotter the capacitor becomes, further increasing the leakage, and so on. Therefore, the operating temperatures of electrolytics must never be allowed to exceed their maximum limit.

The dc leakage of electrolytics should conform roughly to the values given in Table 3-2. These figures will serve as a guide to the maximum allowable leakage current after continuous application of the rated voltage for a minimum of 500 hours. Thus, a 100-μF 150-Vdcw capacitor should have a leakage current of not more than 1.2 mA after 500 hours of continuous operation.

Table 3-2. Maximum Allowable Leakage Current for Electrolytic Capacitors

Rated Vdcw	Maximum Leakage (mA)
25	.002C + 0.1
50	.004C + 0.1
150	.010C + 0.2
450	.020C + 0.3

C = capacitance in μF

The power factor losses of electrolytics tend to be higher than other types. But since the power factor of these units can be misleading, it is more common to describe the losses in terms of equivalent series resistance (esr). Manufacturers' data show the esr on each value produced. For example, the esr for an etched-plate, 100-μF 150-Vdcw unit, is 2.5 ohms. This rating is given for a frequency of 120 Hz at 25°C.

Electrolytic capacitors are rated at 25°C, and the value is stated as a mean capacitance with a tolerance. The latter is determined by the individual manufacturer and will vary depending on voltage rating. A typical tolerance would be −10% to +50% for a capacitor of over 350 volts. Tolerance can be very important to the service technician. Here it means that a 100-μF capacitor actually can range from 90 to 150 μF. In other words, if you are faced with the necessity of replacing a 125-μF unit, it would be quite satisfactory to use a 100-μF capacitor as long as operating temperatures are held well below maximum. The temperatures encountered in normal home-entertainment equipment will usually not exceed this maximum.

Temperature has a very marked effect on the capacitance of electrolytics. At −55°C, a normal aluminum electrolytic has practically no capacitance, and its power factor will be as high as 50%. When equipment has been exposed to extremely low temperatures, all electrolytics will lose a significant percentage of their normal capacitance. The increased power factor may be sufficient to cause internal heating when power is applied. This, in turn, may or may not result in capacitance recovery.

The expected operating life of an aluminum electrolytic is a function of operating temperature, voltage, and "ripple." "Ripple" is the term applied to the fluctuations in voltage encountered when ac is rectified to dc. Electrolytics are commonly used to filter out these fluctuations, particularly in power supplies. The more ripple encountered, the more work the capacitor must do, and hence the shorter its life. As a result, electrolytics which encounter high ripple currents must have adequate safety margins designed in.

To illustrate the effect of ripple on a capacitor consider the following:

1. Life at 65°C at full rated voltage with *no ripple* is 10,000 hours.
2. By reducing the voltage to 75% and still no ripple, life at 65°C is *doubled* to 20,000 hours.
3. Raising the temperature to 85°C, even though voltage remains at 75%, cuts life to 7500 hours.
4. Introducing a ripple condition at 85°C with voltage remaining at 75%, cuts life further to 1000 hours.

The conclusion is obvious. Electrolytics live longer at lower temperatures, lower voltages, or lower ripples (or combinations of all three).

Electrolytic capacitors appear to have many idiosyncrasies. They will gradually deform (decrease in capacitance) if allowed to remain idle. Extended periods of idleness can be harmful to electrolytics. A sudden surge of full rated voltage may then break down the deformed oxide film and permanently ruin the unit. The principal manifestation of this condition is in terms of leakage current (dcl). A capacitor rated 100 μF @ 150 Vdcw could have a leakage current reading of approximately 1.2 mA while operating in a circuit. But, let's remove it from the circuit and store it at room temperature for six months. If we *then* measure the dcl, it could read 50 mA initially and then gradually drop in about a minute or so to its original reading of 1.2 mA.

In order to reform an electrolytic, you can use a capacitor checker, starting with a low (10% of rated) voltage setting and increasing it up to the rated maximum over a period of about one minute. Or, you can do the same thing with a 5-watt resistor of approximately 1000 ohms connected in series with a voltage source not exceeding, and preferably below, the capacitor rating. When one considers that the list price of an average unit is about $2.50, it seems wasteful not to take proper precautions when installing a new one. This is especially true if the capacitor has been idle for a year or more.

If this procedure seems a waste of time, consider the extra expense of a customer call-back. You will always hear of technicians who ignore this operation. They may have been lucky, just like those who say they don't believe in changing the oil in their car.

In the preceding paragraphs we have discussed *normal* aluminum electrolytics and pointed out advantages and shortcomings. As in all other capacitors, there are exceptions to the rule. In recent years several newer styles of aluminum electrolytics have appeared. These include extended temperature types, stacked foil types that offer exceptionally low impedance and small solid electrolyte types. Familiarity with these new types is a must.

Extended Temperature Types

The electrolyte used in normal electrolytic capacitors is a combination of boric acid, ammonium borate, and glycol. Exact formulations vary depending on intended application but the basic electrolyte is relatively harmless. In order to extend the low temperature down to −55°C and up to as much as 150°C an electrolyte known as DMF is often used. DMF is a flouride compound and can be hazardous to humans. DMF has the capability of entering the body through the pores of the skin. It is not only hazardous, it can be *lethal*. Therefore, don't let your curiosity get the better of your common sense. *Do not* dissassemble extended temperature aluminum electrolyics and handle them with bare hands. They can kill you.

On the other hand, great care has gone into the manufacture of these capacitors and they are completely safe when properly handled. The fact is they can perform electrically where other electrolytics are completely useless.

Stacked Foil Types

Stacked foil capacitors are generally used in high capacitance types in association with energy storage circuits. They offer dramatic improvements in impedance since each piece of anode foil and cathode connector is attached to a terminal wire. The resulting stack is then housed in a suitable metal container as large as 3 inches in diameter and 5¾ inches high. In addition to the decreased impedance, the equivalent series resistance is far less than wound types. There are no particular hazards connected with this type capacitor other than the fact that they pack so much power into such small packages. Always treat these capacitors with the respect they deserve. Be sure they are completely discharged before touching the terminals with anything other than a suitable resistor.

Solid Electrolyte Types

As we shall presently discover the solid electrolyte tantalum electrolytic capacitor offers some outstanding electrical advantages. Because of this, a great deal of development work has gone into solid electrolyte aluminum capacitors. The reason is simple. Aluminum is inherently less expensive than tantalum and, technically at least, offers the promise of superior performance. Thus far the aluminum type has not enjoyed far flung acceptance. The cost savings promised have not materialized and there has been no dramatic size reduction. This is, however, a capacitor for the future and may yet become popular. The reader is urged to become acquainted.

A final note about aluminum electrolytics. Large "computer grade" capacitors are potentially *lethal* handle them as though the were a bomb—that is, very carefully.

TANTALUM ELECTROLYTIC CAPACITORS

Tantalum electrolytic capacitors best illustrate the axiom that the best performance at a reasonable price will ultimately win. Originally tantalum electrolytics were used only in the highest grade military applications. But, as their advantages became better known and their applications became broader, the price levels dropped dramatically. This is particularly true of the solid electrolyte type of device.

Tantalum pentoxide as a dielectric is even more effective as an insulator than is aluminum oxide. Therefore, a tantalum electrolytic capacitor is substantially smaller than its aluminum counterpart. This makes them particularly attractive for miniature electronic equipment despite their higher equivalent cost. This high cost is due partially to the fact that tantalum is a relatively rare metal. In fact, its name is derived from "tantalize" because it is so difficult to isolate and process. On the credit side, it is highly resistant to corrosion and has a high melting temperature.

The same principles that apply to aluminum electrolytics apply to tantalums. They are available in both polarized and nonpolarized foil form, as well as a unique configuration which utilizes a sintered anode.

Fig. 3-29 shows some representative tantalum capacitors. They range in size from a grain of rice up to a normal metal-can aluminum electrolytic.

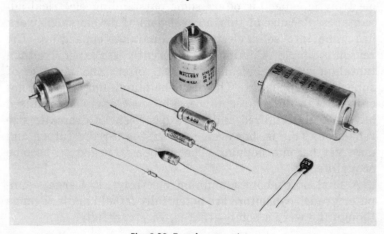

Fig. 3-29. Tantalum capacitors.

In addition to their small size per unit of capacitance, tantalums can be operated at temperatures from −80°C to +200°C. Furthermore, their capacitance is significantly less affected by temperature than aluminum types. Their stability is excellent, and they can be stored almost indefinitely.

This fact alone has led to their adoption by public utilities because of the freedom from reforming them periodically while in storage. Operating life is equally long if normal precautions are observed.

Tantalum capacitors range in value from 0.25 to 2200 μF. Voltages range from 3 Vdcw up to 640 Vdcw. The foil types are normally furnished in axial-lead styles, and may be either polarized or nonpolarized. The sintered-anode types are furnished in a wide range of styles. The electrolyte may be either "wet" or "solid."

Tantalum-foil capacitors are constructed in almost the same manner as the aluminum electrolytic type. The tantalum electrolyte may be sulphuric acid or lithium chloride, depending on the characteristics desired.

There are two basic types of tantalum-foil capacitors. The original type used plain tantalum foil. Tantalum oxide was then formed on the surface to become the anode. Continued research led to the development of a process for etching the foil. Just as in the aluminum type, etching produces a dramatic improvement in capacitance versus size. Etched tantalum-foil capacitors are substantially smaller than plain-foil equivalents. For example, a 20-μF, 150-Vdcw plain-foil tantalum has a case size of ⅜-inch diameter, 2¾-inches long. In contrast, the etched-foil type has 36 μF at the same voltage in the same case. That's nearly twice the capacitance.

Fig. 3-30 illustrates the basic construction of the sintered-anode type of tantalum capacitor. Pure tantalum is produced in minute pellet form. It is then pressed, under very high pressure, into a cylindrical shape onto a supporting and conducting tantalum-wire lead. This cylinder is then fired in a furnace to fuse the areas of contact. It is then electrochemically treated to form a tantalum-oxide coating over the entire spongelike surface. The acid electrolyte is then forced into the cylinder under high vacuum, and the cylinder is assembled into the case, which also serves as the cathode connection. More electrolyte is introduced, and the assembly is sealed.

This type of construction has several advantages. The actual surface area of the spongelike cylinder is very large in proportion to its volume. The electrolyte intimately contacts every part of the oxide film, which in turn contacts every surface of the tiny tantalum pellets. Resistance to physical shock is excellent because the anode is constructed as a single homogeneous unit.

Actual manufacturing practices are much more refined than the drawing in Fig. 3-30 would seem to indicate. Double cases and true glass-to-metal hermetic seals are often employed.

The construction just explained is called the "wet-slug" process. The electrolyte is a liquid, and the sealing problems

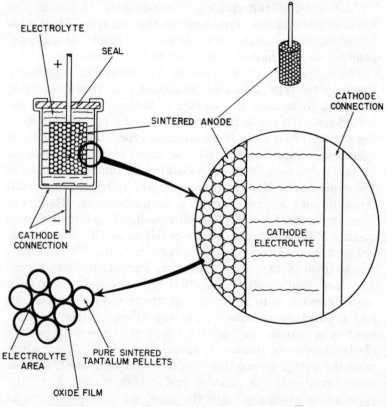

Fig. 3-30. Basic sintered-anode tantalum capacitor construction.

are concerned primarily with preventing the escape of this highly corrosive electrolyte. Operating temperatures range from −55°C to +200°C.

Another type of construction utilizes a solid electrolyte. Its principal advantage is improved performance at very low temperatures. Depending on case structure, these capacitors may be operated at temperatures ranging from $-80°C$ to $+125°C$.

The cylindrical anode is prepared in much the same manner as the wet type, up to the introduction of the electrolyte. At this point, a solid electrolytic material is vacuum-impregnated into the anode and baked to ensure complete dryness and adhesion. Its appearance externally is similar, but the performance is different. Another form of this capacitor uses a flat anode with an epoxy encapsulation and parallel leads similar in appearance to disc-ceramic types. It is useful in printed-circuit applications. The sealing problems of solid-electrolyte types are more concerned with preventing the entrance of outside contaminants and are thus less severe than for the wet type.

Solid tantalum capacitors are found in every type of electronic device from the most exotic sattelite to the cheapest transistor radio. The reason is simple. The very small size of these devices coupled to their high reliability and long life at a very reasonable cost make them very attractive. There is, however, a marked difference in construction between the highly reliable military types and the low cost commercial brands. This is not to say that the low cost types are not good. Rather it is to say that the higher cost military types are good beyond reproach.

Military types of the CSR-13 style made to conform to MIL-C-39003 are among the most reliable devices known to the electronics industry. Their commercial counterparts in the form of dipped or molded plastic case types actually are only small degrees less reliable and at dramatic cost reductions.

Because the solid-electrolyte tantalum capacitor is essentially a monolithic structure, it is relatively unaffected by extreme shocks which could damage tantalums with a liquid or gel electrolyte. On the other hand, certain new liquid-electrolyte tantalums have up to three times as much capacitance as solid-electrolyte types for the same case size. Thus, the designer must decide on which is more important, shock resistance or size.

Just a few short years ago the primary justification for using a tantalum capacitor was its small size coupled with excellent reliability. A typical computer logic board might contain over a thousand capacitors and another thousand precision resistors. The integrated circuit has changed that scene forever. Today one sees only a handful of tantalums on a circuit board. Yet the usage overall of these devices continues virtually unchanged. The reason is, of course, that the IC has expanded the use of electronics by quantum leaps and the tantalum capacitor has found its way into vastly more uses.

CHAPTER 4

CAPACITOR APPLICATION

Selecting the proper capacitor for a specific application is a problem which confronts every electronics technician or engineer at some time or other. In the case of equipment repair, replacement of a defective unit with an exact duplicate is a simple matter. As often happens, however, an exact replacement may not be available, so a suitable substitute must be used. In this chapter, we will discuss some of the important considerations in selecting correct capacitor types for different applications.

OPERATING CHARACTERISTICS

The principal use of a capacitor, from a theoretical standpoint, is for energy storage. In any capacitative circuit some resistance will always be present, offering an opposition to the charge and discharge of the capacitor. Therefore, the RC time constant becomes important and must be considered when one is selecting a capacitor for a particular circuit.

Two capacitors having equal capacitances but different internal resistances will have different charge-discharge time cycles—the unit with the higher resistance taking longer to charge or discharge. This can be expressed by the formula $t = RC$, where t is the time in seconds, R is the resistance in ohms, and C is the capacitance in farads. Since time is a factor in any ac voltage, the RC time constant is directly affected by the voltage frequency. Hence, a capacitor might be usable in a low-frequency circuit, but be unsuitable at higher frequencies. Size, weight, and cost often govern the selection of a particular unit. Therefore, you may encounter a situation when a tantalum capacitor would seem to be the perfect choice and yet find that an ordinary aluminum electrolytic has been used instead. The reason here is the cost. On the other hand, in another circuit, an oil-filled paper capacitor might be electrically indicated, but because of bulk and weight considerations the designer selected an electrolytic.

Yet, despite these particular exceptions, you will find that certain types of capacitors are nearly always used for particular applications. The reason is simple—most engineers are faced with the same problems, consult the same books, and arrive at similar answers. Table 4-1 shows the primary applications of the most common types of capacitors. Glancing over this table, you will note that filtering can be done by almost any type—the overall requirements of a particular circuit determine the final selection.

The preceding paragraphs may have led to a certain amount of confusion in your mind. To partially dispel this, let us now examine some specific applications and find out which capacitors are most suitable.

ENERGY STORAGE

The simplest capacitor application is for energy storage. Here the only problem is storing the required quantity of electrons to meet a specific need. Examples of such applications are welding and electronic photoflash. Primary considerations center on the total energy required, space available, charge-discharge time, and cost.

Table 4-1. Primary Applications of Common Types of Capacitors

	Air	Mica	Ceramic	Paper	Plastic Film	Alum. Elect.	Tant. Elect.
Energy Storage				●	●	●	●
Filtering		●	●	●	●	●	●
Tuning	●	●	●				
Coupling	●	●	●	●	●		
Bypass		●	●	●	●	●	●
Buffer		●	●	●			
Motor Start						●	
Motor Run				●			
Temp. Comp.			●				

We will start with the energy requirement. Assume we are concerned with an electronic-flash application calling for an energy storage of 100 watt-seconds (joules). Our power supply is 450 Vdc. What value of capacitance will be required? The following formula will enable us to make an approximate choice:

$$W = \frac{CE^2}{2}$$

$$100 = \frac{C \times 450^2}{2}$$

$$200 = C \times 202{,}500$$

$$C = \frac{200}{202{,}500}$$

$$C = 0.000987 \text{ farad, or } 987 \ \mu F$$

where,
  W is the energy in joules (watt-seconds),
  C is the capacitance in farads,
  E is the applied voltage in volts.

The fact that the power supply is rated at 450 Vdc and the capacitance is large, means that we should consider using an electrolytic capacitor. We might select a single 1000-μF, 450-Vdcw unit, or two 500-μF capacitors in parallel.

Suppose, however, that our power supply was 2500 Vdc. Obviously, we could not use an electrolytic capacitor at such a high potential. The only other type available is an oil-filled paper dielectric. Using the formula as before, we find that a value of approximately 32 μF is indicated. The unit is much larger and heavier, but is comparable in cost to the electrolytic.

Another consideration in energy storage is the charge-discharge cycle. If either cycle is fairly long, almost any capacitor will do. The shorter either cycle becomes, however, the more important internal resistance becomes.

You will recall that we considered this cycle in Chapter 2. Typical curves are shown in Fig. 4-1. The length of time a capacitor takes to reach full charge is a function of the

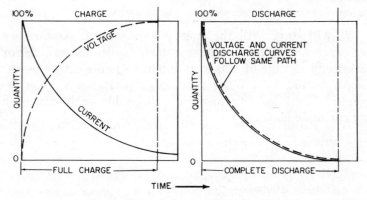

Fig. 4-1. Charge-discharge cycle of a capacitor.

charging force and the internal-capacitor and external-circuit resistances. Therefore, a capacitor with a higher internal resistance will have longer charge and discharge cycles. If the circuit requires very short cycles, we must select suitable capacitors. One method is on the basis of work required. Here we might use much larger units than are needed, in effect shortening the time cycle. This would result in a further advantage—the internal heating of an otherwise unacceptable unit would be reduced to a safe level. The cost would be higher, however.

From the service technician's standpoint, capacitor replacement is much more important than original selection. Also, it is his responsibility to make certain the replacement matches the characteristics of the original equipment.

FILTERING

Capacitors in a rectifier circuit are used principally to smooth out dc pulses derived from an ac source. In this sense, they function as storers of energy. The degree of smoothness required and the inherent circuit characteristics are used as the basis for capacitor selection.

Below 450 volts and/or 1000 Hz, electrolytics are an excellent choice if current requirements are fairly high. At higher voltages and/or frequencies, paper or plastic-film types will be required. Still higher voltages and/or frequencies will require micas or ceramics.

As is true for pure energy storage, the RC time constant of filter capacitors is of prime consideration. It must match the requirement of the overall circuit, particularly if you are replacing a mica capacitor with a ceramic type. (Naturally, all other characteristics must match in this instance, too.)

The job of a filter capacitor is to remove all unwanted frequencies. Consider an antenna filter, for example. One can select a series of capacitors and coils which will eliminate virtually all but the important range of signals needed. Air-dielectric types are the most common here because of the high frequencies and very low currents. An added feature is that tuning to other frequencies is easier.

Another use of filter capacitors is in power-supply circuits, as shown in Fig. 4-2A. Here the essential function is to filter out the unwanted ac variations in order to produce an essentially ripple-free dc voltage. In this circuit the input to the filter capacitors is a pulsating dc voltage with a frequency of 120 Hz. At the first impulse, capacitor C1 starts charging through the rather low resistance of the trans-

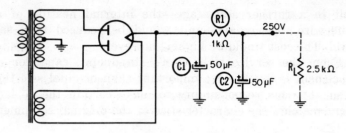

(A) Full-wave tube rectifier.

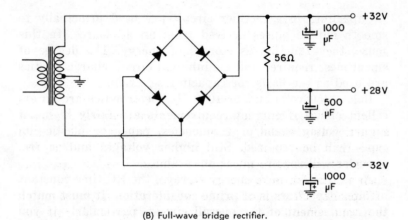

(B) Full-wave bridge rectifier.

Fig. 4-2. Typical power-supply circuits illustrating the use of capacitors as filters.

former winding and the conducting rectifier tube. Because of its large capacitance, C1 takes several cycles to become charged completely, however. During this period C2 is also attempting to charge, but through a slightly higher resistance since R1 is in series with it. After the first few cycles, both capacitors will be charged to the supply voltage. With no load (R_L disconnected), the voltage at the output terminals will remain constant since there is no external discharge path for the two capacitors. If the load is now connected, a path will be provided, but the RC time constant of the circuit will be so long that the filters will have only a very short time (4000 microseconds at 120 Hz) to discharge. Even with such a short discharge time, the output voltage will drop slightly, but will rise on the next charge cycle. This produces a slight ripple with a resultant hum in the output. As long as the ripple is low in amplitude, however, the hum will not be bothersome.

The circuit shown in Fig. 4-2B is similar to the tube circuit shown in Fig. 4-2A except that a full-wave bridge rectifier now replaces the vacuum tube. If the equipment uses tubes in the circuit, there is virtually no change in the values of the filter capacitors. However, the ripple current will be less, because of the more efficient action of the rectifier.

On the other hand, if the equipment is a solid-state circuit utilizing diodes and transistors, the output voltage will be much less than 250 volts. In this case, the values of the filter capacitors will be much higher (on the order of 1000 μF) at lower voltage ratings (perhaps 25 or 50 Vdcw).

FREQUENCY BYPASSING

It is often desirable, and sometimes necessary, to eliminate a particular frequency by bypassing it to ground. Where more than one frequency must be passed and others bypassed, other components are needed as well as the capacitor. Such a situation is shown in Fig. 4-3.

Capacitor C7 and variable resistor R2 function together as a tone control in a typical radio receiver. Note that the two are connected in series from the plate of the af amplifier to ground. Audio frequencies are present at the plate and

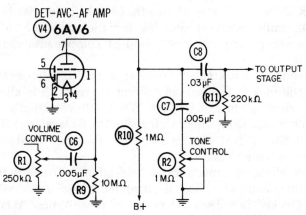

Fig. 4-3. Circuit illustrating frequency bypassing action of an RC combination.

appear across R11 for further amplification in the output stage.

R2 and C7 form an RC circuit, the total impedance of which is determined by the resistance of R2 and the capacitive reactance of C7. At 1000 Hz, C7 has a reactance of approximately 32,000 ohms; while at 10,000 Hz, the reactance is only about 3200 ohms. When R2 is adjusted for maximum resistance, the total impedance of the tone-control circuit will always be greater than 1,000,000 ohms, which is sufficiently high that any audio frequency signal appearing at the plate of V4 will be relatively unaffected and will be passed on to the output stage for further amplification.

However, when R2 is adjusted for zero resistance, only the capacitive reactance of C7 is left in the circuit. Since this reactance is low for the higher frequencies, the higher frequencies will be reduced in amplitude much more than the lower frequencies and can thus be said to be "bypassed to ground."

COUPLING OR BLOCKING

Besides their pure energy-storage capabilities and charge-discharge cycles (RC time constant), capacitors possess a number of other characteristics which can be either a detriment or an asset, depending on the intended use. In other words, some of the very things that might otherwise be considered faults make it possible for the capacitor to do a job no other component or group of components could do as well.

Consider a blocking or coupling capacitor, for example. Here the idea is to block any direct current while allowing an alternating current to pass. What we are actually doing is using a capacitor which charges to the dc level, but simply doesn't have a fast enough reaction (RC time constant) to charge at the ac rate.

Most blocking capacitors in television or radio sets are paper or plastic film. Cost is paramount, and the dc voltage to be blocked is within the range of these types. A typical circuit is shown in Fig. 4-4.

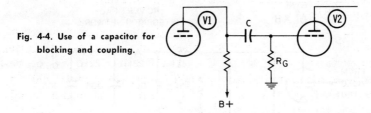

Fig. 4-4. Use of a capacitor for blocking and coupling.

Capacitor C keeps the dc voltage in the plate circuit from appearing on the grid of V2. In other words, C is "blocking" this voltage. At the same time we want any ac signal voltage at the plate of V1 to be transferred to the grid of V2. This can be done through the combined action of C and R_g. Hence, the capacitor has effectively "coupled" the two stages, as far as the signal is concerned.

Here again the time constant formed by the two units must be considered. The RC time must be sufficiently long to prevent the capacitor from appreciably charging and discharging at any audio frequency we might want coupled to the grid of V2. This will cause most of the signal voltage to appear across R_g. If made too long, however, a positive voltage may be built up on the grid, causing distortion and possible tube failure. A compromise must therefore be made. The capacitor is kept near or below 0.05 μF, and resistor values are selected to give a time constant of about 5000 microseconds.

FREQUENCY SEPARATION

Rather than bypassing certain frequencies, it is often necessary to separate them. A practical demonstration of this principle is illustrated in Fig. 4-5. The name *sync separator* is most descriptive of the circuit action, as we shall see.

Both horizontal- and vertical-sync pulses are present at the plate of V4B. Their frequencies differ widely—15,750 Hz for the horizontal but only 60 Hz for the vertical pulses. The RC integrator network offers a higher impedance path to the 15,750-Hz pulses than does capacitor C46. Therefore, the larger portion of the horizontal-sync signal will pass

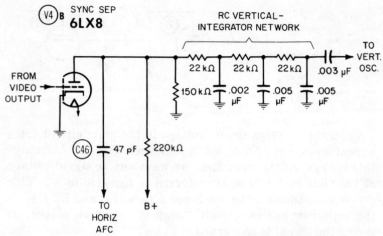

Fig. 4-5. Sync-separator circuit used in a typical television receiver.

through C46 to the horizontal afc circuit, and only a negligible amount will enter the RC network.

To the low-frequency (60-Hz) vertical pulses, the reverse is true—the network offers a much lower impedance path than does C46, so most of the vertical-sync pulses will pass through the RC network and into the vertical-oscillator circuit.

Here again, the choice of capacitor is determined by the characteristics desired and by the current and frequency requirements.

BUFFERS

Buffers utilize the energy-absorption characteristic of a capacitor. They are commonly associated with vibrators or choppers, their purpose being to avoid contact arcing; the same problem exists with switches and relays.

As electrons flow through the closed contacts of a relay or switch, the current reaches some maximum value and remains there. When the contacts are opened, a spark will occur, and if it is sufficiently intense or occurs repeatedly, it can damage the contact material. If the abruptly interrupted current has been flowing through an inductive circuit (a relay coil, transformer winding, etc.), the collapsing magnetic field will generate a high voltage in the inductance, and this high voltage will maintain and even intensify the arc.

Arcing can be minimized by placing a capacitor in parallel with the contacts. The excess electrons that formerly caused the arc are now used to charge the capacitor and thereby prevent damage to the points. As the contacts close, the capacitor discharges, making it ready for the next voltage surge. Fig. 4-6 illustrates a method of contact protection.

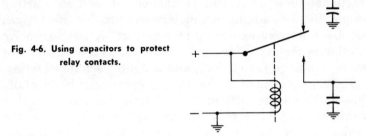

Fig. 4-6. Using capacitors to protect relay contacts.

The most common type of buffer used is the paper capacitor with a value chosen that will be a function of circuit requirements. In some cases, a plastic-film unit might be suitable, but the superior surge-voltage ratings of the paper type make it more desirable.

One problem facing a buffer capacitor is the sudden surge of voltage as the contacts close. Furthermore, rapidly vibrating contacts cause a cumulative voltage buildup in the circuit. The buffer capacitor must therefore be capable of withstanding high voltages and it must resist corona. For this reason, paper capacitors are commonly used as buffers. However, where capacitance is ample, a ceramic type may actually be preferable.

TUNING

There are other applications where the choice of a particular capacitor type is so clear that very little discussion is required. Such is the case in tuning. The air-dielectric type is predominant here, because radio frequencies are involved, stability is very important, and long mechanical life is desired. Ceramic or mica trimmer capacitors may be used to complement the air unit.

Not all tuning devices are capacitors, however; a great many are inductive units. They will not be discussed in this book, except to note their existence and to point out that they cannot be replaced directly by capacitive devices.

AC-MOTOR STARTING AND RUNNING

For many household appliances, it is desirable to have fractional-horsepower motors capable of a high starting torque. Repulsion-induction motors will provide this torque, but are relatively expensive to construct. A less expensive solution is the capacitor-start motor. It has a normal set of heavy windings for running and a lighter set for starting.

The capacitor used in the starting cycle must be carefully chosen. This is because of the RC time constant. In other words, the capacitor must charge fast enough to store the needed energy. Yet, it must discharge fast enough to avoid bucking the next cycle of current. In addition, rather large amounts of capacitance are required to deliver enough energy to get the motor started. The most economical capacitor is the electrolytic type. Since it will be connected directly across the ac line, it must be of the nonpolarized construction.

A centrifugal switch is employed which connects the motor-start windings to the capacitor. As running speed is reached, the switch disconnects the capacitor, and the motor continues as a pure single-phase unit.

Motor-starting capacitors are now commonly located in a housing on the outside of the motor. This improves the cooling of the capacitor and allows the motor to be more compact. Older styles had the capacitor mounted inside the motor housing. In case one of these older types needs a capacitor replacement, it is perfectly acceptable to mount the capacitor on the outside of the frame. Mounting brackets are available from all capacitor manufacturers for this purpose.

Major causes of capacitor failure include aging and excessive start cycles, voltages, and temperature. Another major cause of failure is mechanical—if the centrifugal-switch contacts fail to open, the capacitor will quickly fail. On the other hand, if they open too soon, running speed will not be reached and the start cycle will occur repeatedly until the capacitor eventually fails.

As previously noted in Chapter 3, electrolytic capacitors are seriously affected by low temperatures. Thus capacitor-start motors exposed to extremely low temperatures run the risk of relatively short capacitor lives. An interesting situation is the home central air conditioner. This generally consists of the compressor and heat exchanger being mounted outside of the home. For much of the northern United States, these units are subjected to extremely low temperatures during the period of time when they are unused in the winter. This combination of low temperature and several months of disuse can take its toll of motor-start capacitors. As a result, there is inevitably a brisk replacement business in the late spring and early summer.

Many motor-start capacitors incorporate a 15,000-ohm, 2-watt resistor connected directly across the capacitor terminals. This resistor is used to bleed any residual charge away after the centrifugal-switch points open. Otherwise, the stored energy would tend to buck the first phases on a restart.

Before replacing a motor-start capacitor, always check the mechanical condition of the motor for bad bearings, dirty switch, etc., first. It is always best to mount the replacement on the motor exterior even though the original was not. In this way, the capacitor will run cooler and last longer.

Motor-run capacitors are used in conjunction with a low starting-torque motor like the one in a fan. Here, quietness of operation is the primary advantage and its small size a close second. Motor-run capacitors are usually oil/paper types. Failures are traceable primarily to excessive voltage and/or temperature.

A third type of motor is the capacitor start and run. It is used in large fans and in air conditioners, where both high starting torque and quiet operation are required. This motor is merely a combination of the two mentioned previously.

CHAPTER 5

CAPACITOR REPLACEMENT

There are certain precautions you should observe before replacing a capacitor. The mere fact that it has the correct value and voltage rating does not guarantee successful results.

Suppose that you are called on to repair a defective television receiver and you find that two mica capacitors have failed. Unfortunately you have no exact replacement micas with you, but you do have ceramics. Can they be used instead? The answer is a qualified yes—you can almost always replace a mica with a ceramic of equivalent value. There is one thing to be careful of, however, and that is the tolerance rating.

TOLERANCE

Capacitor tolerance is expressed in percent. Let's take the case of a $\pm 10\%$ unit rated at 10 pF. This means that the manufacturer guarantees this particular capacitor to have a value not less than 9 nor more than 11 pF. But remember, he only guarantees this at the rated voltage and at 25°C

(77°F). If the temperature is higher, the capacitance will also be higher, and vice versa. This is true in all types except ceramics. The capacitance-temperature characteristics of ceramics do not follow the rules for the other types—they are engineered into the specific dielectric used. You'll have to make certain these characteristics fit the job at hand. (You'll find much of the information you will need in Chapter 3.)

Next, let's look at the tolerance for a typical electrolytic. A tolerance of −10% to +50% means that a 100-μF unit will have an actual capacitance value of between 90 and 150 μF. In other words, you will have to measure the capacitor to find its exact value. So remember—don't use ordinary electrolytics where capacitance values are critical.

Tolerance is a measure of the overall quality of a capacitor. It is obviously more expensive to manufacture a capacitor with a very close tolerance than with a very wide one. The most practical method of accomplishing this, from the manufacturer's standpoint, is to build a group of units to a certain specification that should result in a certain value, and then test and label them for their specific value. As the voltage rating of an electrolytic increases, the capacitance tolerance becomes less. A typical range would show capacitors rated at 1 to 50 Vdcw and having a tolerance of −10% to +250%; above 350 Vdcw the tolerance would be reduced to −10% to +50%. Data on specific values are available from every capacitor manufacturer and most of the time from your electronic parts distributor.

Ceramic capacitors are rated either in percent or in guaranteed minimum value (GMV). The percentage was explained earlier. The GMV is simply the manufacturer's guarantee that the unit will have no less capacitance than is stated. (Remember, this is at 25°C.)

TEMPERATURE

Let's return to the problem of substituting ceramics for micas. In addition to being careful about the tolerance, you will have to make certain that the ceramic capacitor has the same electrical characteristics as the mica unit. In other words, you certainly wouldn't want to use a negative temperature-compensating type if it would upset the circuit. The safest thing to do would be to use an NPO type; it has a flat temperature curve.

Let's look at the other side of the coin for a moment. Suppose the capacitor which has gone bad is a ceramic, and all you have are micas. If the bad ceramic is one of the general-purpose types, go right ahead and put in the mica, because the circuit requirements are probably not too critical. But if the ceramic is one of the temperature-compensating types, you're in trouble—a mica won't do here. Its temperature-versus-capacitance characteristic just doesn't match that of the temperature-compensating ceramic.

TYPE CONSIDERATION

Here's another problem. Can you replace a paper capacitor with one of the newer plastic-film types. For most applications the answer is, "Yes, but with some exceptions." For example, you probably would not wish to replace a paper buffer capacitor with a plastic-film type. This is because of the high voltages encountered in buffer applications. Paper capacitors have superior corona characteristics. However, many ceramic types may be substituted for paper types in buffer use.

The foregoing does not mean that paper capacitors are superior to film types. Far from it. Indeed, many dual-dielectric film capacitors (paper/film) are vastly superior to the pure paper type. Therefore, by making an intelligent choice you can usually use the newer film type. They are generally smaller and much more reliable.

Now you might ask, "Can I replace a film capacitor with a paper type?" The answer is, "Maybe." The paper type will probably be larger, but that's only of minor consequence. The entire circuit could be upset because the designer of the equipment carefully chose the original capacitor for its characteristics.

Many inexpensive radios, and even some television receivers, use wax- or paraffin-filled paper capacitors. They are more subject to moisture infiltration than the molded-case types. For this reason, the latter should be used when replacement is required. As a matter of fact, if you run into one or more bad paraffin-filled capacitors in a circuit, it's usually a good idea to check the rest of them while you're at it.

Getting back to the replacement of micas with ceramics, and vice versa, you may run into a situation where a tubular ceramic needs replacement. There is no problem here—you can nearly always replace a tubular ceramic with an equivalent disc type. As a rule, the small difference in inductance will have no effect. However, this is not true if the tubular ceramic happens to be a feedthrough type—the presence of a feedthrough ceramic practically guarantees that an inductance problem exists. The only thing to do here is use an exact replacement.

A similar situation occurs with trimmer capacitors—if the trimmer is an air type, replace it with another air type. If the trimmer is a mica, you can consider a ceramic, provided their overall characteristics match. But be careful about substituting a mica trimmer for a ceramic trimmer —the temperature characteristics of the two may differ radically.

ELECTROLYTIC SUBSTITUTION

Can you replace an electrolytic with an equivalent paper type, and vice versa? If the value of the electrolytic is low enough, you might be able to find a paper type that would fit the available space, although it's highly unlikely. Remember, the electrolytic was no doubt chosen in the first place because of its smaller size. Also, it may have some peculiar electrical characteristic. Replacing it with a paper unit—even of the same value—could be a serious error.

The converse is even more true. Except in the case of pure energy storage (dc), a paper type cannot be replaced with an electrolytic, even with one of equivalent value. Paper types were chosen originally because of their superior voltage characteristics, or because of polarity-reversal considerations. Remember, polarity is vitally important in an electrolytic capacitor. Even a momentary reversal could ruin one. Also, electrolytics do not have as close a tolerance as paper types do.

VOLTAGE RATING

Service technicians encounter all sorts of problems in a day's work. Consider this one, for example. A 0.02-μF, 200-Vdcw paper capacitor has shorted and there is no exact replacement anywhere in the shop. What would you do? You could use a 400- or even a 600-Vdcw unit and no one would ever know the difference. Even a 1000-Vdcw unit might be all right, although it will be slightly larger and cost a bit more. Then, too, the electrical characteristics will start to change, because the voltage rating of a capacitor is determined by the thickness of its dielectric. Therefore, in order to gain the same capacitance, the unit must have a larger plate area. This in turn may change the internal resistance. If the circuit requirements are not too critical, a 1000-Vdcw unit may be satisfactory. On the other hand, it may introduce other problems.

As a general rule, it is safe to substitute a capacitor of a higher voltage rating for one of a lower voltage rating. But don't try to go the other way—if a circuit requires a 600-Vdcw capacitor and you use a 400-Vdcw unit, you can expect more than your share of trouble.

PARALLEL CAPACITORS

Another fairly simple problem is one where a certain capacitor is called for, but only smaller values are available. For example, suppose you need an 80-μF, 300-Vdcw capacitor—simply place two 40-μF, 300-Vdcw units (or any other reasonable combination) in parallel. This brings up the possibility of upsetting the RC time constant. The fact that the internal resistances are also paralleled now results in a lower overall resistance. Although you can parallel capacitors to increase capacitance, you cannot increase the voltage rating in this manner. The voltage rating remains the same because, unlike capacitance, voltages in parallel are not additive.

SERIES CAPACITORS

There is a stop-gap method of increasing the voltage rating, and that is to place capacitors in series. Thus, two 40-μF, 150-Vdcw capacitors in series will be equal to a single 20-μF, 300-Vdcw unit. Understand that in both series and parallel cases the RC time constant may be upset, since the resistance is lower in the parallel circuit and higher in the series circuit than it would be in an equivalent single capacitor.

SHAPE CONSIDERATIONS

The shape of a capacitor is often an important consideration due to space requirements. Substitution of one shape for another may radically change the distributed capacitance. For example, it may be perfectly satisfactory to substitute a disc for a tubular ceramic. Likewise, a molded *Mylar* capacitor that has axial leads might be replaced with another that has radial ones. Trouble might develop from this substitution in critical circuits, however, due to changes in the distributed capacitance.

Shape and size are also important in electrolytic replacement, especially in compact equipment where the space occupied by the defective unit does not permit substitution of a differently shaped capacitor. The two most common types of electrolytics are the tubular and the metal can, and their performance and internal construction are identical. Hence, as long as their values are the same, they may be freely interchanged as far as electrical requirements are concerned.

MULTIPLE UNITS

Very often, two, three, or four electrolytic capacitors are furnished in a single can. In instances where only one of them fails, a satisfactory repair can usually be made by disconnecting the defective portion and soldering a single tubular in its place. This is assuming, of course, that the remaining sections have not suffered any damage due to failure of the defective section (such as from overheating). Also, if the known bad section has failed due to deterioration which has taken place over a period of time, the other sections are more subject to early failure. In such cases, it would be wise to replace the entire unit.

When an exact replacement for a four-unit (quad) electrolytic is not on hand—a triple-unit electrolytic and a single capacitor together work just as well.

REPLACEMENT TECHNIQUES

Removing the defective capacitor and inserting the replacement is seemingly a simple operation. Yet here is where many mistakes are made.

Faulty installation can affect circuit performance even though the new unit is a perfect twin. Consider the mica capacitor in Fig. 5-1. Originally it was installed at right angles to the chassis, as in Fig. 5-1A. The replacement has been installed parallel with and too close to the chassis (Fig. 5-1B). A possible shift in capacitance is likely because of the difference in the distributed capacitance (Fig. 5-1C) between the metal chassis and the plates inside the capacitor.

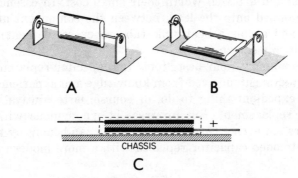

Fig. 5-1. Faulty installation of a capacitor producing undesirable coupling to chassis.

Lead Length

Another common error is failure to recognize the effect of lead length on the resonant frequency of capacitors and their circuits. For example, changing the length from 0.5 to 0.3 inch can raise the resonant frequency of a disc ceramic by as much as 10 MHz. Other types may be affected even more.

Disc ceramics are singled out because they are so common in miniature circuitry, and when components are tightly packed there is always the temptation to make the replacement the easy way. Thus, in a critical circuit make certain that the replacement leads are the same length as those in the original.

Soldering Precautions

Miniaturization has led to a new set of problems for the service technician—all components are smaller, and capacitors are no exception. Consider the small *Mylar* types, for example. Naturally they must be soldered into the circuit. But the heat of the soldering iron can ruin them unless proper precautions are taken. Always use a soldering device that provides no more heat than is needed. Prolonged application of a hot iron to the lead of a *Mylar* unit can melt the dielectric or the internal solder joint between lead and plate. A deft hand is the surest way to avoid this problem.

An excellent device for preventing heat damage is a set of surgical clamps. They are sold under a variety of trade names, and are well worth their small cost. In essence they are clamped onto the lead, between the component and the iron, and act as a heat sink (absorb excessive heat). This principle is shown in Fig. 5-2.

In summary, your best "tool" for capacitor replacement is good judgment, derived from knowledge and experience. The most expedient thing to do, of course, is to always use an exact replacement. But progress is synonymous with electronics . . . even a three-year old radio can benefit by having an outmoded capacitor replaced with a more modern one.

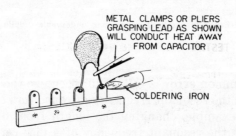

Fig. 5-2. Protecting capacitor from excessive heat during soldering.

CHAPTER 6

CAPACITOR TESTING

How can you know that a capacitor is bad? Or for that matter, how will you know that it is good?

Beyond the fact that a capacitor lead is completely burned off or broken in two, there are other ways a capacitor can become defective. It may be shorted or completely open (have no capacitance at all). The capacitance may have drifted from the desired value, or the internal resistance may be too low or too high. The dc leakage may also be much too high. Most of these defects can be detected fairly easily.

A few tests can be performed with the unit still in the circuit; others require that at least one lead be unsoldered.

TESTING WITH THE VOM OR EVM

In complex circuitry, particularly where printed boards and/or miniaturized components are used, it is best to make as many in-circuit tests as possible. There's no sense in unsoldering or removing components from the circuit until other tests lead you to believe that such action is necessary.

Consider, for example, the coupling capacitor, C8, in Fig. 6-1. If a voltage measurement indicates that a potential of 25 volts is present at the grid of V2, it is reasonable to assume that the capacitor could be leaky. But before carrying this assumption further, why not obtain additional proof? After all, a shorted V2 could be responsible for the positive grid voltage. Unless there is reason not to do so, the tube can be removed from the socket and the grid voltage measured again. If the same condition is noted, and there seems to be no other contributing factor, it's time to disconnect the grid end of the capacitor from the circuit.

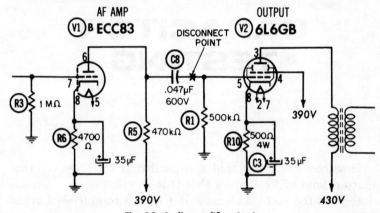

Fig. 6-1. Audio amplifier circuit.

As shown in Fig. 6-2, there are two methods of testing for capacitor leakage, using nothing more than a vom or evm. With the set turned on, and no signal coming through, you need only measure voltage or current by connecting the appropriate meter between the open end of the capacitor and ground. Normal readings, depending on capacitor type and value, should not exceed 2 or 3 volts, or 5 to 10 mA. A word of caution in making these voltage and current measurements: A badly shorted capacitor will permit high currents to flow through the ground-return path provided by the meter, so be sure to start out on a scale high enough to prevent damage to the meter.

There is one other simple, although not entirely conclusive, in-circuit test you can make for a shorted capacitor. If a

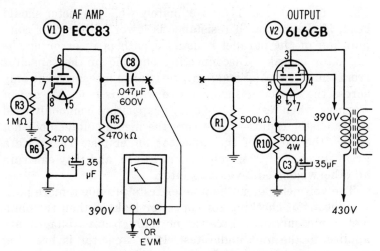

Fig. 6-2. Use of vom or evm for testing capacitor leakage.

capacitor is badly shorted, you can usually discover the defect with an ohmmeter. *Before making this test, make sure no voltages are present in the equipment; otherwise, damage to the meter will result.*

As shown in Fig. 6-3, merely connect the ohmmeter across the suspected unit, using the scale that provides a usable reading. In order for the test to have any validity, you must compute the resistance value for existing parallel paths.

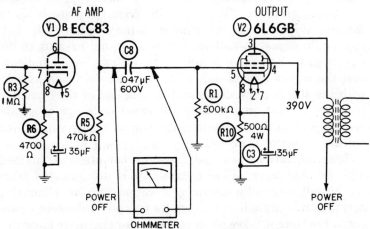

Fig. 6-3. Use of ohmmeter to detect shorted capacitor.

This computed value is approximately what the meter should read. If the measured resistance is lower, either your computation of the parallel resistance path is in error or the capacitor is leaky. Disconnecting one end of the capacitor from the circuit and making a resistance measurement only across the capacitor should prove which is true. A word of warning when testing low-voltage capacitors commonly used in transistor radios—make sure the meter voltage does not exceed the rating of the capacitor under test. There are a great number of 22½-volt meters in use, and they can ruin an otherwise good 3-volt capacitor.

The voltmeter or milliammeter tests provide a more positive means of checking for capacitor leakage than the ohmmeter measurement because normal circuit voltages are applied to the unit under test. If a capacitor is breaking down intermittently, it is more likely to do so when B+ voltages are applied; the ohmmeter measurement does not take this factor into account.

Checking for open capacitors is a fairly simple task also. But again, by making in-circuit tests first, you can save much time and trouble. One of the best tests involves the use of an oscilloscope or ac voltmeter to check for the presence of signal. Of course, you must know what to expect in the way of waveform or voltage indications. For instance, you would expect a coupling capacitor such as the one in Fig. 6-1 to pass practically all of the signal available from the plate of V1 to the grid of V2. On the other hand, in the circuit of Fig. 6-4, you would expect to find very little ac signal present at the screen grid. Excessive signal at this point would indicate that the screen bypass unit (C4) was not doing its job. As with the shorts tests, disconnecting one end of the capacitor from the circuit, and making the same tests again, should provide sufficient information for you to decide whether or not the capacitor is defective.

Some indication of an open capacitor can often be obtained with an ohmmeter. When a capacitor is disconnected from the circuit and fully discharged, connecting an ohmmeter across it will cause it to charge. As was pointed out in Chapter 2, the time it takes for a capacitor to charge is governed by the value of capacitance and the amount of series resis-

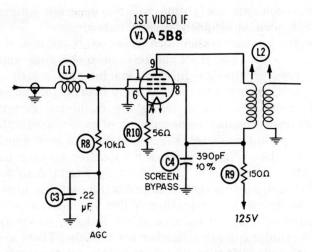

Fig. 6-4. Television video i-f stage showing screen bypass capacitor.

tance in the circuit. The ohmmeter itself has several thousand ohms of resistance (depending on the scale used) ; thus, if the capacitor is of fairly large value—say, 0.01 μF or more —the time constant will be such that it can be distinctly noticeable from the change in the ohmmeter reading. On the 100-kΩ scale, for example, the initial reading for a normal capacitor may be in the neighborhood of 500 kΩ. That is, the pointer may swing over as far as midscale, and then swing back—rapidly at first, and then gradually more slowly—until the reading is near infinity. If you could accurately plot the resistance readings on a time scale, you would find they duplicate the capacitor charging curves discussed in Chapter 2.

As you have undoubtedly surmised by now, failure of the ohmmeter to react as described is a pretty good indication that the capacitor is open. Remember, however, that the capacitance value must be sufficient to provide a measurable charging period. A value of 330 pF might not produce more than a slight flicker of the ohmmeter needle. Incidentally, the phenomenon known as dielectric absorption, described in Chapter 2, is most likely to be encountered when using the highest scale of the ohmmeter (usually 1 meg or 10 meg). Therefore, to obtain a valid test, switch to the next lowest scale. (Dielectric absorption will not generally be encountered in mica, or electrolytic capacitors.)

As far as the capacitance value is concerned, a vom equipped with an ac scale can be used to obtain an approximate reading. Specific instructions for this measurement are usually included in the operating manual for the instrument. Some instrument scales are calibrated directly in microfarads, making measurement of capacitance values a relatively simple matter. Fig. 6-5 illustrates such a unit.

Using the ordinary vom as a capacitor checker has its limitations. In the first place, it really wasn't designed for the job and, for this reason, is primarily useful in providing only preliminary information. When it comes to accurate measurements of capacitance value, leakage, power factor, etc., a regular capacitor checker is called for. There are two general types—the in-circuit and the out-of-circuit—and each has its particular applications and limitations.

IN-CIRCUIT CAPACITOR CHECKERS

The in-circuit capacitor checker will tell you at a glance whether or not a capacitor is shorted or open. But it cannot

Fig. 6-5. A vom-capacity tester.

Courtesy Mercury Electronics Corp.

tell you the exact value of a particular capacitor. It is, in essence, an all-or-nothing device. Most of the time this will do—as far as you're concerned, either a capacitor is working or it is not. Fig. 6-6 shows an in-circuit checker.

Courtesy Mercury Electronics Corp.

Fig. 6-6. An in-circuit capacitor tester.

OUT-OF-CIRCUIT CAPACITOR CHECKERS

There are circuits where the value of the capacitor can be especially critical. For instance, in a divider network a slight shift in capacitance can seriously affect performance. If trouble is indicated here, the capacitor must be removed from the circuit (or, at least one lead unsoldered) before it can be accurately checked. All out-of-circuit capacitor checkers, such as the two shown in Fig. 6-7, use some modification of a balanced-bridge circuit. The principle involved is simple —a known capacitance is balanced against the unknown capacitance by coupling an appropriate variable resistance to a calibrated dial. When the needle shows that the circuit is balanced, the capacitance value can be read directly from the dial. Other checkers may use variations of this circuit,

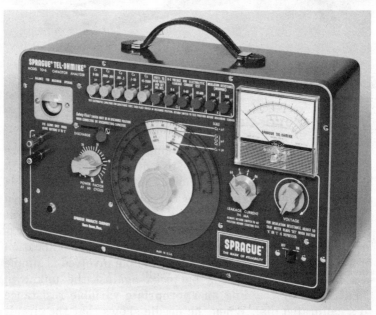

Fig. 6-7. Two typical out-of-circuit capacitor analyzers. These units can measure capacitance values, leakage current, power factor, and resistance.

but they accomplish the same results. A basic bridge circuit, used in many out-of-circuit testers, is shown in Fig. 6-8.

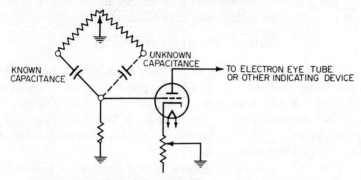

Fig. 6-8. Bridge circuit used to measure value of unknown capacitance.

The leakage resistance of capacitors varies from very high in micas and ceramics to very low in electrolytics. Checking for a change in this characteristic can be performed only with an out-of-circuit tester or an appropriate bridge. Since values range up to 20,000 megohms, the correct test setup is required to ensure accuracy.

Excessive dc leakage is a common problem with capacitors. It is usually expressed in terms of the internal resistance, because leakage is low whenever resistance is high. In electrolytics, the leakage is high enough that it can be compared directly. The test for dc or ac leakage must be performed with the capacitor out of the circuit. Manufacturers of both test equipment and capacitors supply charts indicating the correct amount for each type and value. These should be used in place of any rule-of-thumb, and even then, good judgment will be necessary. It is essential to keep the capacitor at the temperature specified (usually +25°C) when checking for dc leakage; otherwise, you will obtain an inaccurate reading. Usually, ac leakage is expressed as so many milliamps at a particular frequency (such as 120 Hz) and temperature (such as +85°C).

Out-of-circuit testers are also useful for measuring the power factors of electrolytics. A bridge similar to the one in Fig. 6-8 is used. Here again, the test must be performed at rated temperatures only.

Another method of checking capacitance, described in Chapter 2, is to use the time constant. It is generally restricted to elaborate laboratory tests, mainly because a great deal of expensive equipment is involved.

DIGITAL AUTORANGING TESTERS

The advent of the large scale integrated circuit has made possible an entirely new type of capacitor tester. This new tester is not only vastly more accurate than older types, it is also a breeze to use. All one need do is attach two leads from the tester to the capacitor and read the capacitance directly on the digital meter.

Until 1975 capacitors testers of this class were priced in the thousands of dollars. Then a new device came on the market. It is known as the ECD "C" Meter (Fig. 6-9) and

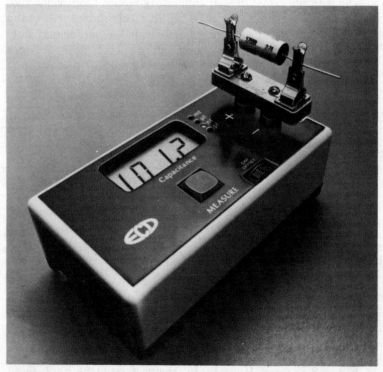

Fig. 6-9. Digital autoranging capacitor tester.

sells for less than $300.00 (1977). It is powered by four AA batteries and is totally portable. It measures capacitance by the time constant method as described in Chapter 2. The difference being that the use of LSI integrated circuits permits the meter to do the job precisely and in fractions of a second.

The meter is autoranging from 1.0 picofarad up to 199,900 microfarads and has an accuracy of 0.1% to 200 μF and 1% from 200 to 200,000 μF. This accuracy is maintained within the temperature range of 15 to 35 degrees C (59—95°F). Four LED indicators show the range of capacitance being scaled and take just a bit of getting used to. This is because millifarads and nanofarads are somewhat strange at first.

Index

A

Ac-motor, starting and running, 98-99
Air dielectric capacitors, 38-40
  fixed, 38
  variable, 38
Aluminum electrolytic capacitors, 73-81
Anode
  capacitor, 10, 37
  electrolytic capacitors, 73-75
  sintered, 84-85
Arcing, prevention of, 96-97

B

Battery, definition of, 8
Blocking capacitors, 94-95
Bridge circuit
  balanced, use for capacitance measurement, 115, 117
  method of capacitance measurement, 24
Buffer capacitors, 96-97

C

Capacitance
  basic formula for, 14

Capacitance—cont
  basic units of, 12-13
  effect of temperature on, 20-21
  formula for measurement of, 25
  formula for series, 42
  measurement of, 12-13, 24-26
    balanced-bridge circuit, 115, 117
    time-constant method, 25-26
  temperature-coefficient of, 21
Capacitive reactance, 31-33
  definition of, 31
  formula for, 32
Capacitor
  air-dielectric, 38-40
  anode, 10
  application, 87-99
    ac-motor, starting and running, 98-99
    arcing, prevention of, 96-97
    blocking, 94-95
    buffer, 96-97
    coupling, 94-95
    energy storage, 87-91
    filtering, 91-93
    frequency bypassing, 93-94
    frequency separation, 95-96
    relay-contact protection, 96-97
    tuning, 39, 97-98

Phenolic resin, 45
Picofarad, definition of, 12
Plastic-film capacitors, 37, 57-62
   cellulose triacetate, 61
   polycarbonate, 61
   polyester, 57-58
   polyester/polystyrene, 60
   polyparaxylene, 61
   polypropylene, 61
   polypyromellitimide, 61
   polystyrene, 59
   polytetrafluoroethylene, 61
Plate losses, capacitor, 23
Plates
   capacitor, 11
   rotor, 39
Polarized electrolytic capacitors, 72
Polycarbonate capacitors, 61
Polyester-film capacitor, 57-58
Polyester/polystyrene capacitors, 60
Polyparaxylene, 61
Polypropylene, 61
Polypyromellitimide, 61
Polystyrene film capacitor, 59
Polysulfone, 62
Polytetrafluoroethylene, 61
Power factor
   capacitor, 19-20
   dielectric material, 22
   electrolytic capacitors, 78
PTFE, 62
P-type ceramic capacitors, 64

R

RC time constant, 29-31, 87-88
Reconstituted mica capacitor, 49
Relay-contact protection, 96-97
Replacement, capacitor, 101-108
Residual charge, capacitor, 30
Resin, phenolic, 45
Resonance, 34
Resonant frequency, capacitor, 19
Ripple, electrolytic capacitor, 79
Rotor plates, 39

S

Safety, capacitor, 15-16
Sawtooth generator, neon-lamp, 30-31
Self-healing capacitor, 56
Semiconductor material, 37
Semipolarized electrolytic capacitors, 73
Series capacitance, formula for, 42
Series capacitors, replacement considerations, 106
Shape considerations, capacitor replacement, 106
Shorted capacitor, testing for, 110-112
Silver-mica capacitors, 45
   dipped, 47
Silver migration, 71
Sintered anode, tantalum electrolytic capacitors, 84
Soldering precautions, capacitor replacement techniques, 108
Solid electrolytic capacitor, 81
Stacked foil capacitor, 81
Sync separator, 95-96

T

Tab-conductor construction, 54
Tantalum electrolytic capacitors, 81-86
   construction of, 83-85
   effect of temperature on, 82-83
   electrolyte, 83
      liquid, 84
      solid, 85
   etched foil, 83
   plain foil, 83
   sintered anode, 82, 83-84
   uses of, 81-82
   "wet-slug" process, 84
Tantalum pentoxide, 22
Techniques, capacitor replacement, 107-108